International Conference on Emerging Trends in Science and Circuit Engineering (ICESCE 2021)

IOP Conference Series: Materials Science and Engineering
Volume 1177

Davangere, India
23-24 April 2021

Editors:

K. B. Bhaskar
A. Paramasivam

ISBN: 978-1-7138-4637-6
ISSN: 1757-8981

Printed from e-media with permission by:

Curran Associates, Inc.
57 Morehouse Lane
Red Hook, NY 12571

CURRAN ASSOCIATES INC. proceedings.com

Some format issues inherent in the e-media version may also appear in this print version.

Institute of Physics
Dirac House, Temple Back
Bristol BS1 6BE UK

Phone: 44 1 17 929 7481
Fax: 44 1 17 920 0979

techtracking@iop.org

Additional copies of this publication are available from:

Curran Associates, Inc.
57 Morehouse Lane
Red Hook, NY 12571 USA
Phone: 845-758-0400
Fax: 845-758-2633
Email: curran@proceedings.com
Web: www.proceedings.com

TABLE OF CONTENTS

IOP Conf. Series: Materials Science and Engineering 1177 (2021) 011001 doi:10.1088/1757-899X/1177/1/011001

Preparation of editorial material to appear in the proceedings

• List of possible items for the editorial section of the Work:

o Preface/introduction (required)

This is the first International Conference on Emerging Trends in Science and Circuit Engineering 2021. ICESCE – 2021 aims to bring together the scientists, researchers, engineers, industrial professionals and scholar students in their domain of interest for the dissemination of novel research results, new ideas, practical development experiences which concentrate on both theory and realistic around the world in all the areas of Science, Power Systems, Power Electronics and Drives, Energy, Robotics, Embedded Systems, Electric Vehicles, Battery Systems & Control and Instrumentation Engineering.

o List of committees (required)

Chief Patron

- Dr. S. Dilli Saran raj, Chairman

Patron

- Smt. Sujatha Maran, CEO

Co-Patron

- Dr.A.Devaraju, Principal

Convener

- Dr.K.B.Bhaskar, Associate Professor/ Head of the Department

Editorial Board

- Dr.K.B.Bhaskar, Head of the Department, Electrical and Electronics Engineering, Adhi College of Engineering and Technology, No.6, Munu Adhi Nagar, Sankarapuram, Walajabad, Kanchipuram – 631 605, Tamilnadu, India. Contact No: 9786851178 Email: eeehod@adhi.edu.in
- Dr.A.Paramasivam, Assistant Professor, B.S. Abdur Rahman Crescent Institute of Science and technology, GST Rad, Vandalur, Chennai – 600 048, Tamilnadu, India. Contact No: 044 – 22751347, Email: paramasivam.eee@crescent.education

Executive Committee

- Mr.R.Kumaresan, Assistant Professor

- Mr.M.Rajavelu, Assistant Professor
- Ms.D.Jasmine Susila, Assistant Professor
- Mr.R.Aravind, Assistant Professor
- Ms.E.Kokila, Assistant Professor
- Mr.M.Dinesh, Assistant Professor
- Mr.S.Rajesh, Assistant Professor
- Mr.N.Jehan, Assistant Professor

Advisory Committee

- Dr.Vignakumaran, Professor, University Tenega Nasional, Malaysia.
- Dr. Lenin Gopal, Associate Professor, Curtan University, Malaysia.
- Dr.T.S.Sivakumaran, Professor, Bule Hora University, Oramia, Ethiopia.
- Dr.Vicente Rodolfo Garcia Colon Hernandez, Researcher at National Institute of Electricity and Clean Energy, Mexico.
- Dr.B.Chitti Babu, Assistant Professor, IIITDM, Kanchipuram.
- Dr.P.Somasundaram, Professor, CEG, Anna University, Chennai.
- Dr.J.Baskaran, Principal, ARM College of Engineering and Technology, Chennai.
- Mr.P.Ramanavasu, Additional Chief Engineer, IGCAR, Kalpakkam.
- Dr.V.S.Bharath, Professor, Oxford Engineering College, Bangalore.
- Dr.K.Ravi, Professor, VIT University, Vellore.
- Dr.R.Ramesh, Associate Professor, CMR Institute of Technology, Bangalore.
- Dr.S.Jeevananthan, Professor, Pondicherry Engineering College, Pondicherry.
- Dr.C.Bharathi Raja, Associate Professor, SRM Institute of Science and Technology, Katankulathur.
- Dr. J. Balamurugan, Assistant Executive Engineer - TNEB/TANGEDCO, Chennai.
- Dr.R.Femi, Research Assistant Professor, SRM institute of Science and Technology, Katankulathur.
- Mr.S. SelvaKumar, Business Head, Power Projects, Chennai.
- Dr.S.Prakash, Professor, Bharath Institute of Higher Education and Research, Chennai.
- Mr.Gobi Durairaj, Assistant Lead Test Engineer, Vestas Technology R&D Pvt, Ltd, Chennai.

IOP Conf. Series: Materials Science and Engineering 1177 (2021) 011001 doi:10.1088/1757-899X/1177/1/011001

- Dr.S.Vijayalakshmi, Assistant Professor, Vel Tech Rangarajan Dr.Sagunthala R&D Institute of Science and Technology, Chennai.
- Dr.A.Deepak, Electromagnetics Design Engineer, Epropelled Systems Pvt Ltd, Chennai.
- Dr.A.Paramasivam, Assistant Professor, B.S. Abdur Rahman Crescent Institute of Science and technology, Chennai.
- Dr.Hina Fathima, Project Lead, Ford Motors Company Ltd, Chennai.
- Dr.A.Prasanth, Young Scientist & Assistant Professor, PSNA College of Engineering and Technology, Dindigul.

Reviewer Committee

- Dr.D.Palanisamy, Adhi College of Engineering and Technology, palanisamy.mech@adhi.edu.in
- Dr.Sankara Gomathy, Adhi College of Engineering and Technology, sankaragomathi.ece@adhi.edu.in
- Dr.T.S.Sivakumaran, Bule Hora University, Oramia, Ethiopia
- Dr.V.S.Bharath, Oxford Engineering College, Bangalore.
- Dr.R.Ramesh, CMR Institute of Technology, Bangalore
- Dr.R.Femi, SRM institute of Science and Technology, Katankulathur.
- Dr.S.Prakash, Bharath Institute of Higher Education and Research, Chennai.
- Dr.S.Vijayalakshmi, Vel Tech Rangarajan Dr.Sagunthala R&D Institute of Science and Technology, Chennai
- Dr.A.Paramasivam, B.S. Abdur Rahman Crescent Institute of Science and technology, Chennai
- Dr.T.Muthamizhan, Sairam Institute of Technology, muthamizhan.eee@sairamit.edu.in
- Dr.A.Prasanth, Young Scientist & Assistant Professor, PSNA College of Engineering and Technology, Dindigul
- Dr.M.Ettappan, Chennai Institute of Technology, ettappanm@gmail.com
- Dr. Thanga Chelliah, IITR, India, ctr.iitr@gmail.com
- Dr.Ali I. Maswood, NTU, Singapore, mohd0017@e.ntu.edu.sg
- Dr.S.Jeevanathan,PEC,India, drsj_eee@pec.edu
- Dr.B. Hariram, ABB, India satheesh@in.abb.com
- Dr.S.Paramasivam, Danfoss,India, paramsathya@yahoo.com
- Dr. Ramani Kannan, Universiti Teknologi PETRONAS, Malaysia, kreee82@gmail.com

IOP Conf. Series: Materials Science and Engineering 1177 (2021) 011001 doi:10.1088/1757-899X/1177/1/011001

IOP Conf. Series: Materials Science and Engineering 1177 (2021) 011002 doi:10.1088/1757-899X/1177/1/011002

Peer Review Declaration Form

All papers published in this volume of *IOP Conference Series: Materials Science and Engineering* have been peer reviewed through processes administered by the Editors. Reviews were conducted by expert referees to the professional and scientific standards expected of a proceedings journal published by IOP Publishing.

- **Type of peer review:**

  Double-blind

- **Describe criteria used by Reviewers when accepting/declining papers. Was there the opportunity to resubmit articles after revisions?**

  The papers received from authors were sent to reviewers shortly. Based on the reviewer comments, suggestions were recommended to the authors within a week of time. The corrected manuscript from the authors according to the IOP quality, were submitted to IOP.

- **Conference submission management system:**

  icesce2021@gmail.com

- **Number of submissions received:**

  9

- **Number of submissions sent for review:**

  6

- **Number of submissions accepted:**

  5

- **Acceptance Rate (Number of Submissions Accepted / Number of Submissions Received X 100):**

  56%

- **Average number of reviews per paper:**

  2

IOP Conf. Series: Materials Science and Engineering 1177 (2021) 011002 doi:10.1088/1757-899X/1177/1/011002

- **Total number of reviewers involved:**

  12

- **Any additional info on review process (eg Plagiarism check system):**

  Turnitin

- **Contact person for queries (Full name, affiliation, institutional email address)**

Dr.K.B.Bhaskar, Head of the Department, Electrical and Electronics Engineering, Adhi College of Engineering and Technology, No.6, Munu Adhi Nagar, Sankarapuram, Walajabad, Kanchipuram – 631 605, Tamilnadu, India.

Contact No: 9786851178

Email: eeehod@adhi.edu.in

Design of Transformerless Single Stage Double Switch Step Down PLL Derived Inverter using Single Phase Grid Connected PV Module with MPP Tracking

A Vijayaprabhu[1] K B Bhaskar[2] M Rajavelu[3] N jehan[4]
[1]Assistant Professor, Department of Electronics and Communication, Sri Venkateswara College of Engineering and Technology, India
[2,3,4]Department of Electrical and Electronics Engineering, Adhi College of Engineering and Technology, India

Vijayabrabhu85@gmail.com[1], jaibhaskar15@gmail.com[2]

Abstract. Here we propose a single stage double switch (SSDS) converter for single phase grid connected photo voltaic inverter step-down module, the structure is designed and modelled to a highly reliable step-down buck DC-DC converter and robust from the input derived from the PV module. The proposed PV module uses perturb and observe algorithm reference voltage to gain the maximum power in different changing climatic conditions. The grid is designed for 110 V/50 Hz, since it's a low frequency, the utilization of heavy and expensive transformer is avoided, In a grid connected systems usually transformer is employed in the power conversion stage to provide sudden isolation between the PV system and the grid, and also ensures that no continuous current is injected into the grid by avoiding leakage currents between the PV module and the ground. As a choice to the grid connected module with line transformer, a transformer less inverter topology is discussed, to avoid the leakage currents possibly occurring due to the capacitance between the PV array and ground, an inverter topology with PLL is proposed to avoid common mode voltage and the increase in electromagnetic emissions. This paper provides a highly efficient transformer less photovoltaic grid topology with reduced switches and has been validated by simulation.

Keywords: Step-down converter, grid interface, transformer less, PLL, MPP Tracking.

1. Introduction

Striving over the decades, the requirement for fossil fuels have keep on increasing as it cannot fulfil the requirement for day to day energy demand. They are getting compensated by the renewable energy generation systems are connected to the grid in a reliable way [1]. This began to rise in semiconductor power devices in power system.1 phase photo voltaic grid connected inverter systems gains an optimal solution for low frequency systems [2,3].

Considering any one of the PV topology systems from standalone and grid-connected, In the grid connected systems the power semiconductor converter topology plays a vital role in transporting the available photo voltaic dc output in to the grid [4,5]. One more classification of standalone photo voltaic systems required a converter setup to convert produced DC in to AC for the end users. Power semiconductor switching devices are necessary in making interaction with the produced DC output with the AC load applications [6].

The switching device made the converter should always keep in track of photo voltaic module produces maximum VI in spite of the variation in the climatic changes and load variations[7]. This is taken in consideration by MPP tracking, the converted continuous AC voltage fed to the loads from the generated PV modules[8], the produced output must match the parameters of the grid such as alternating nature of output, frequency and amplitude[9].

In power inverters based grid to maintain the level of synchronism in order to produce an instantaneous alternating voltage to provide the VI to the loads, To synchronize the photovoltaic tracking output and the AC grid a PLL (Phase-locked loop) was proposed[10], [11], which performs the angle detection in the grid, the grid stability has to be maintained perfectly therefore the controlling to the inverter should be modified from the grid to the leading grid nature[12], this type of grid-forming arrangement where semiconductor inverters can able to produce AC with correct amplitude and stipulated frequency at the point of common coupling[13].

The paper proposes a **SSDS** 1-phase grid connected inverter, mainly concentrates on voltage source inverter grid-forming control for low frequency applications, the proposed controlling to the PI controller gains improves the stability of the system and approached to operate at unity power factor to enable efficient utilization of the full inverter capacity. Since, operation at any random power factor is possible in cases of availability of flexible inverter capacity due to possible drop in the available solar power, therefore the system adopts the MPP tracking with the incremental conductance method and without difficulty it can track any variations in insulation without oscillation. The system is designed in matlab/Simulink to verify its proposed characteristics.

Up to this, a system elaboration is made in the first section, Further, this paper elaborates as, in the next section II describes the system model, PV modelling with characteristics, MPP tracking methodology, Proposed Double switch step down DC-DC converter module in section III. The output simulation model are discussed, conclusions are done in section [IV, V].

2. Modeling & Tracking

2.1. System Model

The SSDS proposed converter model presents a 1 Phase grid interface photo voltaic model, in the below Figure1 shows the system consists of PV array with MPP tracking fed the input to the proposed step-down converter which will reduce the supply voltage for low level applications and PLL is used for grid side integration to the inverter.

IOP Conf. Series: Materials Science and Engineering 1177 (2021) 012001 doi:10.1088/1757-899X/1177/1/012001

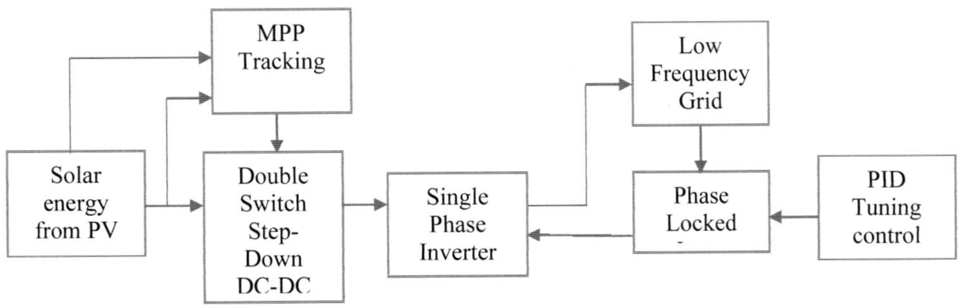

Figure 1. Block diagram of the proposed model

2.2. *PV modelling with characteristics*

The PV cells are arranged in a bunch units are called PV panels which are collectively modelled in a parallel-series to give the PV arrays.The general voltage and current equation of PV array with respect to efficiency is very sensitive to cell series resistance Rs. and irrespective to changes in cell shunt resistance Rsh, the simplified equation is as,

$$I_{pv} = N_p I_{ph} - N_p I_0 \left(e^{\frac{q\left(\frac{V_{pv}}{N_s} + \frac{I_{pv} R_s}{N_{pe}}\right)}{KTA}} - 1\right) \quad (1)$$

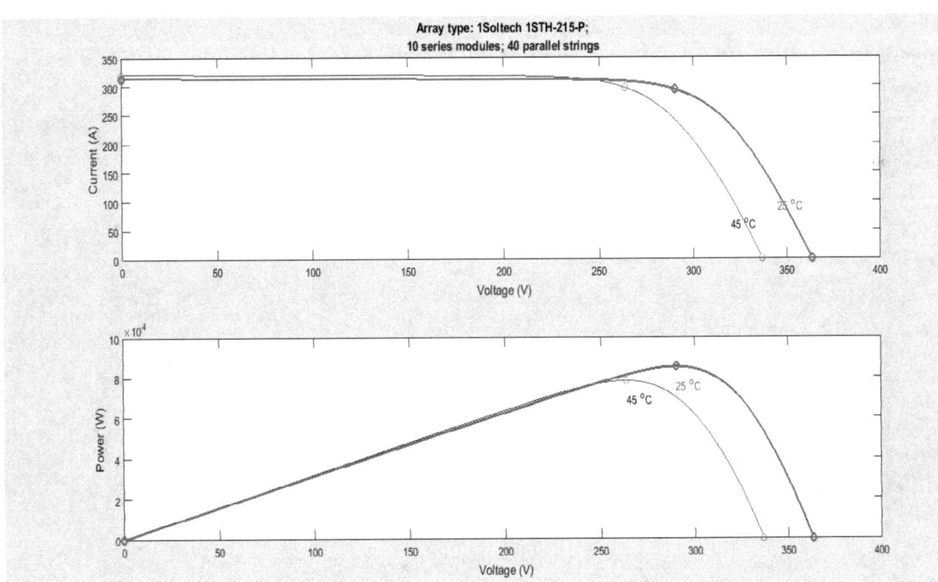

Figure 2. PV Current (A) and Power (W) output Vs voltage (V)

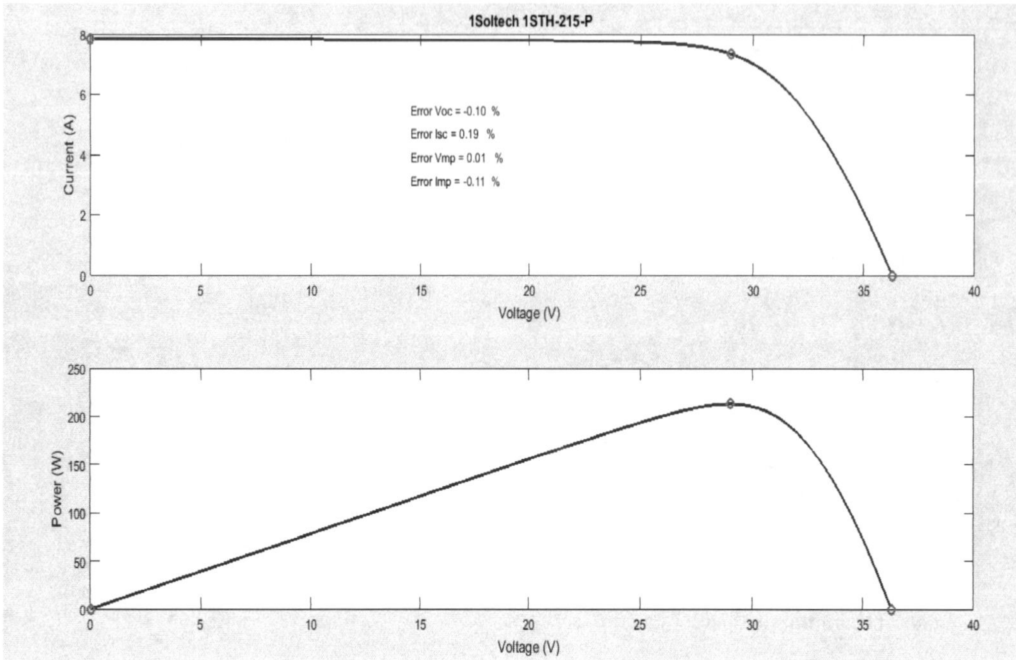

Figure 3. Current (A) error and Power (W) Vs voltage (V)

2.3. *MPP tracking of the Solar panel*

From the photovoltaic modelled in the above equation the maximum power and the output voltage are depend on irradiation and cell temperature to produce a maximal power and maintains the output voltage at a constant value despite of the variations in the temperature a MPP tracking must be used. The MPPT techniques approaches to incremental conductance and P and O process is used and implementation is simple.

Incremental conductance is based as:

$$P_{pv} = V_{pv} \times I_{pv} \text{---------} (2)$$

Maximum power flows from the PV array when,

$$\frac{dp_{pv}}{dv_{pv}} = 0 \text{-------} (3)$$

By applying differentiation to equ (2), we have,

$$\frac{dI_{PV}}{dV_{pv}} = -\frac{I_{pv}}{V_{pv}} \text{------} (4)$$

IOP Conf. Series: Materials Science and Engineering 1177 (2021) 012001 doi:10.1088/1757-899X/1177/1/012001

From the equation (4), the error signal resulting from the inequality is integrated to generate the reference value of the capacitor voltage that minimizes the error signal and ensures the flow of maximum power from the PV array.

3. Proposed SSDS- step down converter

The proposed SSDS step down converter exits in between the power generation and the inverter, the converter topology is buck type unidirectional process flow, the IGBT Q1 and Q2 produces a switching for a certain period based on frequency depends on load current requirement. The inductor 'L' and 'C' acts as a filter and averages the switched voltage.

In the first operating mode shown in Figure , IGBT switch Q1 and Q2 is at ON condition, the capacitor(C) charges through it, and the current flows through the path Vs-Q1-Q2-C The variable pulse width oscillator controls On/OFF periods for Q1 and Q2.

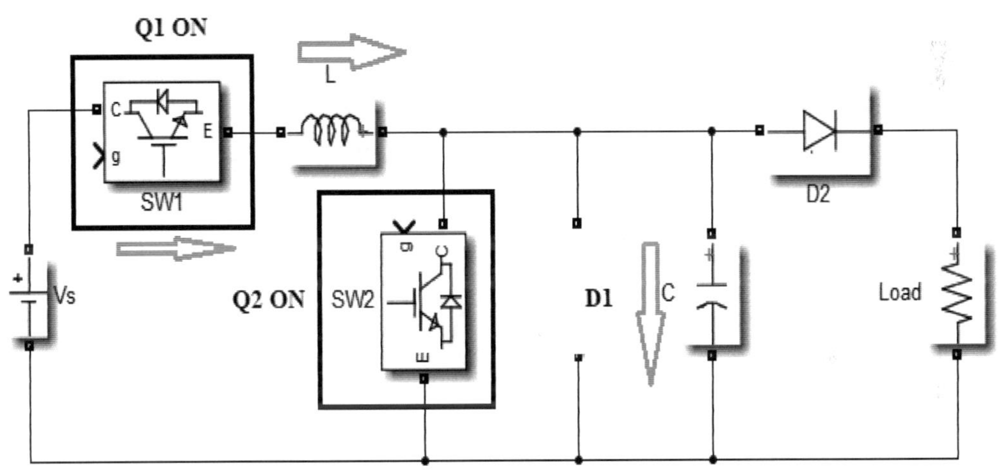

Figure 4. Operating mode 1

In the second operating mode shown in figure, IGBT switch Q1 and Q2 is at OFF condition, the stored energy in the capacitor (c) discharges the energy to the load, the process flow is as L-C-D2-Load.

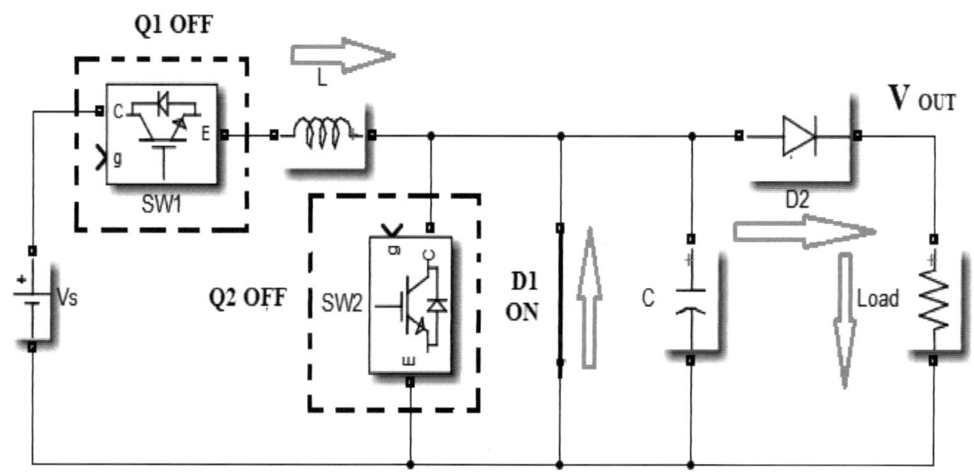

Figure 5. Operating mode 2

When the ON switching time is more compared to OFF time, the capacitor charges more, increasing the output voltage, in parallel when OFF time is more than ON time for Q1 and Q2 the output voltage decreases, here in this case the above mentioned second approach is modelled to produce the buck output from the tracking control to the switches the PWM pulses to the switches is shown in the below Figure 6.

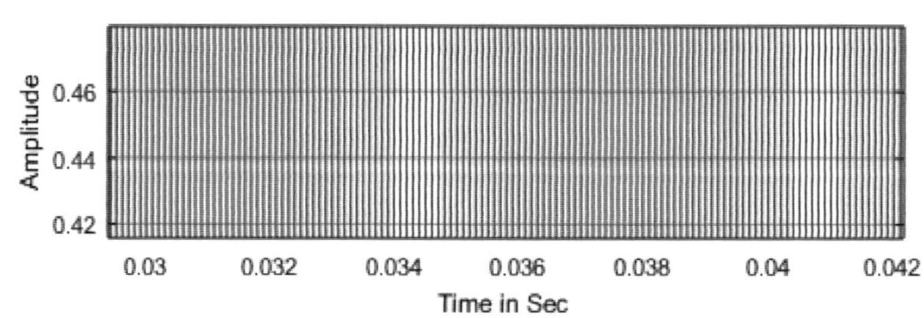

Figure 6. PWM pulses to the switch

4. Simulated circuit design and output waveforms

The proposed single stage double switch in single phase grid connected inverter is simulated using MATLAB/SIMULINK is shown in the Figure 7. The system elaborates the input side PV generation is fed in to the proposed SSDS buck converter and the step down voltage is interfaced with the grid using the inverter topology and the isolation is corrected by PLL technique, and through the inverter AC output is obtained.

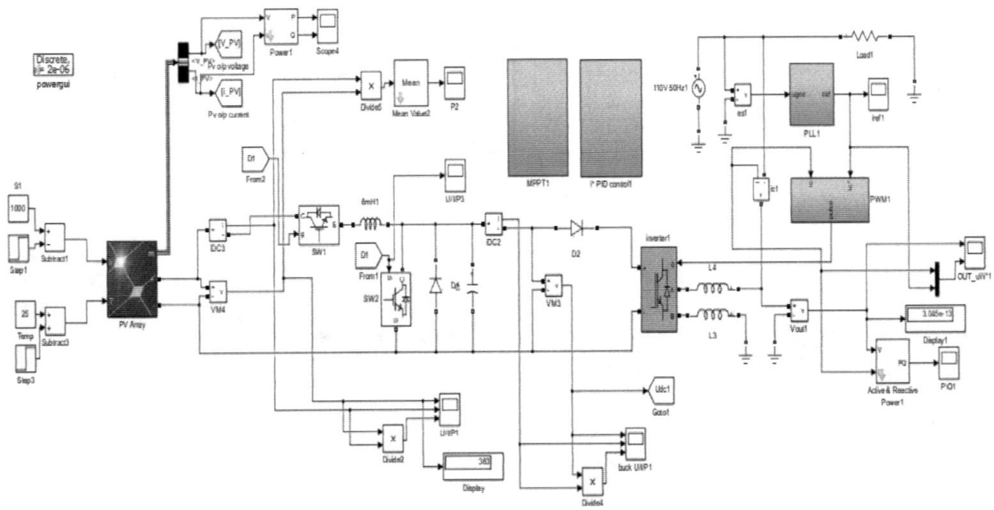

Figure 7. Overall Simulink model for the proposed design

Figure 8. Shows the waveform of the generated power from PV (1Soltech 1STH-215-P) module, Figure 8(i) refer the input voltage of 363V, and 8(ii) & 8(iii) represents the current and power waveforms.

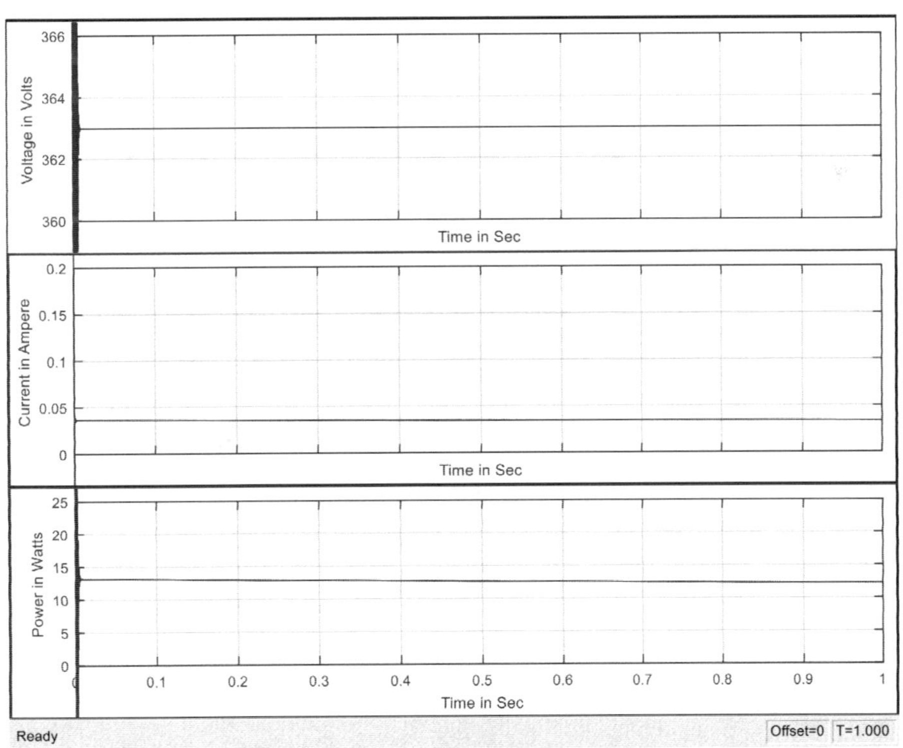

Figure 8. Generated input (i) Voltage (ii) Current (iii) Power

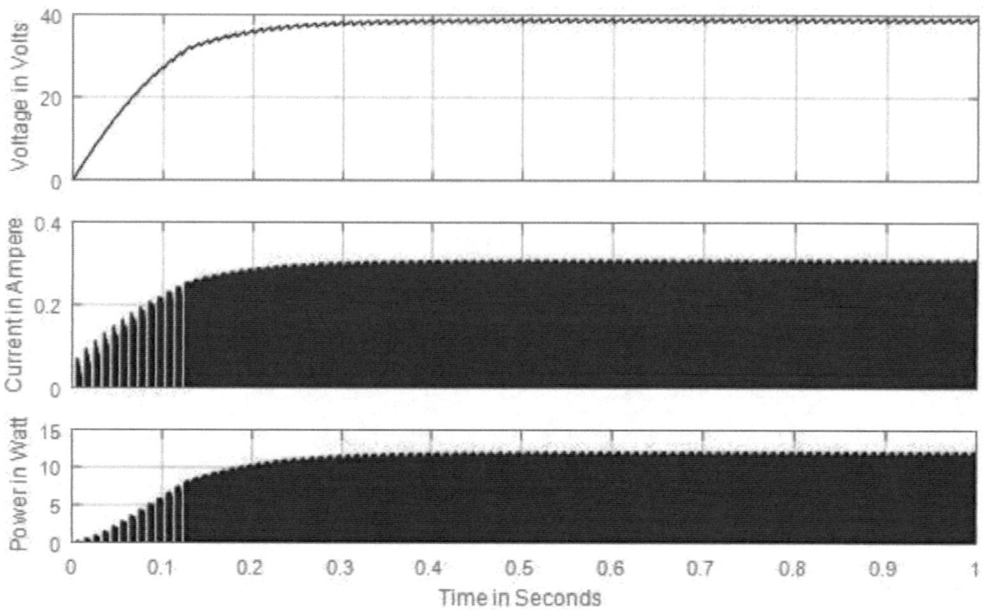

Figure 9. DC Link (i) voltage (ii) Current (iii) Power

Figure 9. Shows the graph of DC capacitor link parameters of PV and Grid side, in that Figure 9. (i) Shows the voltage waveforms which has maintained constant for interfacing to grid, Figure 9. (ii) & Figure 9. (iii) Waveform outputs the link current and power.

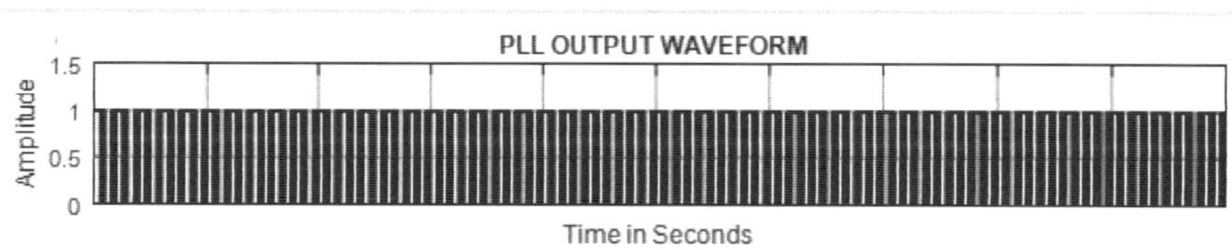

Figure 10. Phase Locked Loop (PLL) pulse

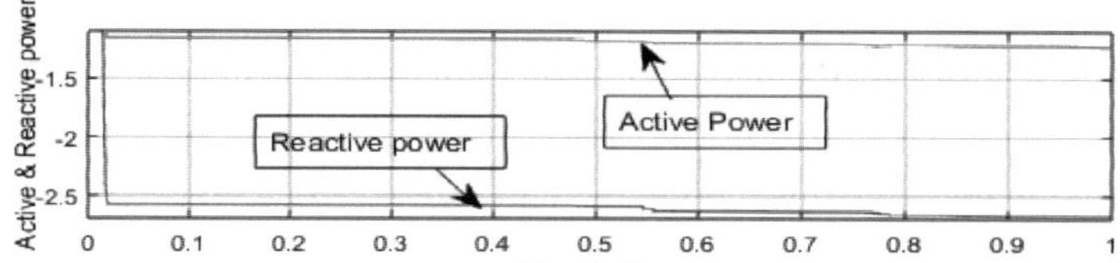

Figure 11. Active power and Reactive power at the output

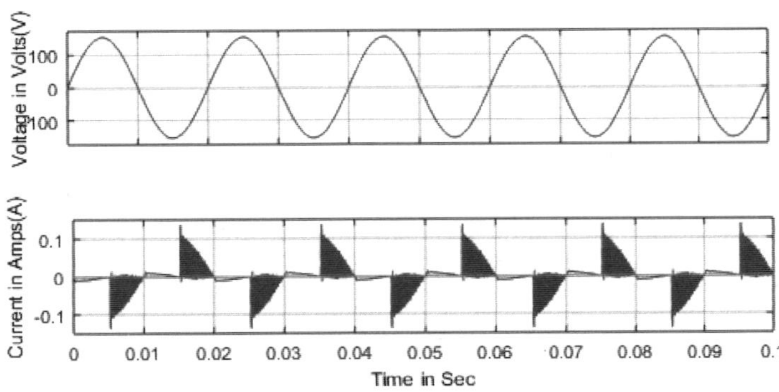

Figure 12. (i) Output Voltage (ii) Output current waveforms

Figure 10. Shows the PLL Pulse waveforms which controls the grid interface inverter and makes synchronization between the generation and grid side, Figure 11. Shows the active and reactive power at the inverter side and Figure 12. (i) Shows the inverter output produced at positive peak of 150V which is a step-down output from the input side, Figure 12. (ii) Inverter output current.

5. Conclusions

In the above proceedings, a single phase grid connected photovoltaic inverter with single stage double switch converter is presented. This system had a non-isolated DC-DC step down and a single phase inverter for DC to AC conversion. The control algorithm of MPP tracking were implemented in order to produce maximum input by performing the controlling of frequency of input switches. The advantage of designing using transformer less circuit avoided the leakage current, In the grid side the Phase Locked Loop (PLL) technique is implemented to provide the synchronization by estimating phase angle for the voltage grid reference. The simulation results shows that the input rated voltage generated by PV is363V and by using the converter the voltage is step-down to 150 V AC at the output end, the output inverter and input solar radiance is always in phase with the grid.

Appendix

PV array parameters

PV Type: 1 Soltech 1STH-215-P

Parallel Strings: 40

Series-connected modules per string: 10

Maximum power: 213.15 W

Cell per module (N cell): 60

Open circuit voltage Voc (V): 36.3 V

Short-circuit current Isc (A): 7.84 A

Voltage at maximum power point Vmp (V): 29 V

Current at maximum power point Imp (A): 7.35 A

Temperature coefficient of Voc (%/deg.C): -0.36099 %/deg.C

Temperature coefficient of Isc (%/deg.C): 0.102 %/deg.C

Model Parameters

Light-generated current IL (A): 7.8649 A

Diode saturation current I0 (A): 2.9259 e-10

Diode ideality factor: 0.98117

Shunt resistance Rsh (ohms): 313.3991 ohms

Series resistance Rs (ohms): 0.39383

References

[1] M. A. Eltawil and Z. Zhao, "Grid-connected photovoltaic power systems: Technical and potential problems-A review," *Renewable and Sustainable Energy Reviews*. 2010, doi: 10.1016/j.rser.2009.07.015.

[2] S. V. Araújo, P. Zacharias, and R. Mallwitz, "Highly efficient single-phase transformerless inverters for grid-connected photovoltaic systems," *IEEE Trans. Ind. Electron.*, 2010, doi: 10.1109/TIE.2009.2037654.

[3] N. F. Guerrero-Rodríguez and A. B. Rey-Boué, "Modelling, simulation and experimental verification for renewable agents connected to a distorted utility grid using a Real-Time Digital Simulation Platform," *Energy Convers. Manag.*, 2014, doi: 10.1016/j.enconman.2014.04.020.

[4] K.B. Bhaskar and T.S.Sivakumaran 'Implimentation of 11 level cascaded multilevel inverter using level shifting pulse with modulation technique with different loads', International journal of electrical engineering(IPASJ),volume 2,issue 10 Nov 2014

[5] D. Debnath and K. Chatterjee, "A buck-boost integrated full bridge inverter for solar photovoltaic based standalone system," 2013, doi: 10.1109/PVSC.2013.6745069.

[6] Bhaskar, K. B., &Alagumariappan, P. (2020). Energy Channeling Led Driver Technology to achieve Flicker Free Operation with Zeta Converter for Power Quality Improvement. *International Journal of Automation and Smart Technology*, *10*(1), 401-407.

[7] P. Vivek, R. Ayshwarya, S. J. Amali, and A. S. N. Sree, "A novel approach on MPPT algorithm for solar panel using buck boost converter," 2016, doi: 10.1109/ICEETS.2016.7583787.

[8] Paramasivam A., Bhaskar K.B., Madhanakkumar N., Vanchinathan C. (2020) Analysis of an Enhanced Positive Output Super-Lift Luo Converter for Renewable Energy Applications. In: Siano P., Jamuna K. (eds) Advances in Smart Grid Technology. Lecture

Notes in Electrical Engineering, vol 687. Springer, Singapore. https://doi.org/10.1007/978-981-15-7245-6_11

[9] Z. M. Yasin, N. Ashida Salim, and N. F. Ab Aziz, "Harmonic Distortion Prediction Model of a Grid -Connected Photovoltaic Using Grey Wolf Optimizer - Least Square Support Vector Machine," 2019, doi: 10.1109/ICPES47639.2019.9105398.

[10] Alagumariappan, Paramasivam, Najumnissa Jamal Dewan, GughanNarasimhanMuthukrishnan, Bhaskar K. Bojji Raju, Ramzan Ali Arshad Bilal, and VijayalakshmiSankaran. 2020. "Intelligent Plant Disease Identification System Using Machine Learning" *Engineering Proceedings* 2, no. 1: 49. https://doi.org/10.3390/ecsa-7-08160

[11] K. Biju and R. Ramchand, "Control of a novel single phase grid connected solar PV/battery hybrid energy system," 2015, doi: 10.1109/ASCC.2015.7244436.

[12] H. Zhao, Q. Zheng, Y. Li, and D. Li, "Study on the Fast Phase-locked Technology Applied in Three-phase Grid-connected System," *Gaodianya Jishu/High Volt. Eng.*, 2018, doi: 10.13336/j.1003-6520.hve.20171227038.

[13] B. Perera, P. Ciufo, and S. Perera, "Advanced point of common coupling voltage controllers for grid-connected solar photovoltaic (PV) systems," *Renew. Energy*, 2016, doi: 10.1016/j.renene.2015.09.028.

Machine Learning Methods for Marine Systems

Dr S Thangalakshmi[1] and Dr K Sivasami[2]

[1]Faculty – Electrical, Indian Maritime University, Chennai Campus
[2]Associate Professor – Marine Engineering, Indian Maritime University, Chennai Campus

E-mail: thangalakshmiprakash@yahoo.com, ksivasami@imu.ac.in

Abstract. Automation plays a key role in shipping industry and aims towards minimal operating staff. However, the effective automation relies on effective controlling at various levels starting from shipbuilding to navigation. The industry is currently focussing on autonomous shipping which actually requires precise controlling. Although many conventional methods are available for control and automation with regard to automation, Artificial Intelligence Schemes (AIS) are widely attracting the maritime sector because of their benefits. The AIS along with fuzzy logic systems are offering promising results. The emerging use of AIS in a variety of maritime applications can act as a reference wpoint for new researchers. This paper aims to conduct a valid AIS study and to examine the various machine learning approaches used in various maritime applications. It is possible to achieve complete automation in the shipping industry by implementing a related technique.

Keywords: Automation, autonomous shipping, artificial intelligence, machine learning

1. Introduction

Presently the digital transformation is one of the inevitable happening in anybody's life and this is quite true after the covid-19 pandemic. While everyone directly or indirectly undergo this digital transformation, a valid and promising technology is needed to accomplish the same. Artificial intelligence systems are one such technique and it applies to almost every industry at a faster rate. Various gadgets and elements incorporate the AIS features in our routine life. These schemes possess altered characteristics based on the device, which adds them or based on a particular application. The approaches and procedures that are followed in seafaring sector are unique and to apply the AIS technology, the distinctiveness of the industry must be appropriately accounted.The scale of the dataset, its authenticity, and the use of recurring key incidents all play a significant role in the success of AIS. This is especially true in the traditional shipping industry, where the transition to full automation is regarded as critical. As a result, a survey was carried out in the most commonly used applications of the maritime industry. The aim of this research is to identify the best area for using AIS and then scale it up to allow full automation. Artificial intelligence can be realized in this field in several facets [1-5].

Machine learning is a subset of artificial intelligence and is a probable way to accomplish AI. Machine learning is capable of dealing with minor to major datasets by means of inspecting and associating the data to get mutual arrays and learn the details. Artificial intelligence is a technology where computers/machines can be trained through a relevant learning algorithm to mimic human brain viz. recognition, remembrance, application, etc. Shipping firms these days have understood that financing in quicker communication offers numerous benefits. Most vessels go forward to remote workplaces at sea, which may offer consistent Internet access, simulated/virtual set-ups, e-mail, route organizers and many supplementary schemes and uses. It is now essential for the shipping firms to plan for long-term growth by investing in novel growing technologies to standardize the vessel operation, CTC reduction and business optimization plans. The shipping sector moves towards automation and AI assists in accomplishing the same [16].

Automated / Unmanned systems do not require rest, they effortlessly do repetitive works, and AI development will deliver work with unstable environments. AI is fundamental for automated systems and effectively replaces the manned systems. AI may perhaps an opportunity for modern shippers. Optimization of maritime operations is of prime importance especially in vessel precision operations. There are various means to precisely make ETA/ETD, perfecting container routing & re-routing, fuel saving, etc. by means of a number of geographical data. These data may accurately be processed using suitable machine learning algorithms [17].

This review was inspired by the promise and prospective of the shipping industry necessitating AIS, as well as the popularity of AIS in other areas. This technique uses data, as an instrument for learning from the bygone events and helps in enhancing the future decision-making. Enhanced metrics for strategy making, scheduling, stability, intelligent traffic, and operational productivity are just a few of the exceptional benefits of AI in the freight sector.

This paper is presented under the following headings: literature analysis, developments in maritime transport, AIS summary, Artificial-Intelligence in marine applications, and conclusion. This paper highlights a few applications/case studies pertaining to AI in shipping industry.

2. Literature Review

The use of machine learning in the shipping industry is very constrained since it analyses vast volumes of data to construct a particular logic. Yet, latest scholars have attempted to apply machine-learning practices in data analysis. Various machine-learning techniques viz. neural networks, Bayesian networks and support vector machines are used to explain the problem of maritime anomaly detection. It is pointed out that the AI data has provided a better level of perception in case of anomaly detection [1]. An enhancement to an existing learning-based maritime field knowledge scheme for forecasting potential vessel position using the experience of current vessel behaviour is discussed, and it is shown that such schemes produce better performance [2]. The AI technique is applied for possible prediction of motion of vessel using learning weights [3]. EnmeiTu et al. [4] surveyed AIS data sources and utilized data for safety of seagoing personnel in the areas of traffic irregularity finding, route assessment, collision forecast and path scheduling. Another literature addressed the trick of determining waypoints of marine routes, using the streaming AIS data [5]. Andrea Coraddu et al. have utilized the machine learning approaches for naval propulsion plants [6]. The various sensors used for effective propulsion and for improving the maintenance based on the condition are aptly administered through predictive algorithms for economic reasons. Artificial intelligence in maritime sector is the current survey by many researchers [7].Deep sea connectivity is one of the most difficult issues in the marine industry. Underwater Wireless Sensor Networks (UWSNs) have seen extensive use in marine and military applications in the recent past. However, in order to outspread the network's survivability in UWSN, effective and truthful fault tolerance mechanisms are essential. The Energy-efficient Fault Detection and Recovery Management approach is recommended by A.Prasanth for the UWSN with reasonably enhanced network efficiency. To achieve effective fault detection across the entire network, the Hidden Poisson Markov model has been introduced into the suggested method. The network simulator (NS 3.26) framework is employed to replicate the performance effects. According to the simulator performance the suggested model outshines the existing evidence [27].

3. Trends in Maritime Transport

Digitalization is the method of integrating digital devices into structures and procedures in order to upgrade or replace traditional devices with electronic devices. The term automation refers to processes and structures, which are mechanical or electronic machines that are routinely computerised to execute specific tasks without human interference but are frequently supervised by humans. The system is considered to be autonomous if it has control tasks and can use various alternatives to solve a certain class of problems [8].

3.1. Degrees of Autonomy for Ships

There exist a variety of definitions of autonomy levels based on the transport used and perspectives for their own systems. The definition of autonomy levels for merchant ships is listed in Table 1. A degree of autonomy is assigned to the system relative to the amount of labour that must be performed by manned operators in conjunction with an automated system.

Lloyd's Register has printed a guide for design of autonomous ships [12]. The guide encompasses a classification of autonomy levels as given in Table 2. Remote monitoring and controlling is not accounted in this taxonomy, except as a choice for shore based decision support [12].

3.2. Infrastructure Operation for Ships

In maritime sector, the infrastructure and control of the vessel operations shift on the way to operation centres that are integrated and interconnected, like fleet service centres, support hubs for marine shipping traffic, etc. These central/integrated activities – maybe for a geologically limited zone like a harbour or for broader capacities like inland waterways – necessitate a technological framework that gives situational information to the controllers that decide things remotely. Such centralized/integrated controlling schemes progress within a broad framework using modern plans to incorporate automated patterns, operating practises, and conventions to ensure maritime vessel protection, effectiveness, and prediction [9].

Table 1. Degree of Autonomy for Ships [8]

Level	Description
Direct control	Management of the ship is exercised directly by the crew on the bridge; no decision assistance is provided.
Decision support	On the bridge, decision support and guidance are given to the crew. The crew makes the final decision.
Automatic bridge	The operation is automated, but the crew is constantly monitoring it.
Periodically unmanned	Shore is constantly monitoring the situation. If required, send in reinforcements.
Remote control	Unmanned, constant surveillance, & command and control from land.
Automatic	Unmanned and under automated supervision, with monitoring from shore.
Constrained autonomous	Unmanned, semi-autonomous, and constantly supervised from shore.
Fully autonomous	It is unmanned and unsupervised.

Table 2. Classification of Autonomy Levels [12]

AL	Name	Decision support on board	Decision support on shore	Automatic, human override
0	Manual	No	No	Human
1	On-ship decision support	Yes	No	Human
2	On and off-ship decision support	Yes	Yes	Human
3	'Active' human in the loop	Yes	Yes	Common
4	Human on the loop	Yes	Yes	When needed
5	Fully autonomous	Yes	Yes	Rare
6	Full autonomous	Yes	Yes	Never

3.3. Infrastructure Operation for Ships
In maritime sector, the infrastructure and control of the vessel operations shift on the way to operation centres that are integrated and interconnected, like fleet service centres, support hubs for marine shipping traffic, etc. These central/integrated activities – maybe for a geologically limited zone like a harbour or for broader capacities like inland waterways – necessitate a technological framework that gives situational information to the controllers that decide things remotely. Such centralized/integrated controlling schemes progress within a broad framework using modern plans to incorporate automated patterns, operating practises, and conventions to ensure maritime vessel protection, effectiveness, and prediction [9].

3.4. Predictive Maintenance of Vehicles
A significant percentage of servicing tasks in marine shipping could be conducted by personnel who are no longer stationed on-board the vessel, but rather in centrally facilitated support centres. Safety checks, such as software upgrades, may be performed remotely from those centres or through highly trained teams stationed in support sites worldwide and sent to the location where maintenance service is requested. Recent advancements in sensor expertise plus genuine networking help to accomplish the same and permit to observe the condition of apparatus on-board vehicles keenly [10].
Future prognostic maintenance schemes will gradually use an extensive collection of large data points. This knowledge is collected not just from vessels and facilities, but rather from peripheral operational configurations that have an effect upon this particular container such as freight, tides, landscape, and climatic parameters along voyages. Predictions can indeed be improved by creating a visual replica of the vessels and their structures (a "digital twin") [11].

3.5. Automation of Vessels
Fresh and creative forms of robotics and automation can support ship crews, spanning from adaptive vehicle tasks, such as watch keeping, velocity monitoring, and propellant usage to automated driving operation, even without manual assistance. Regulation of vessel activities and facilities is moving toward cohesive and integrated operation centres, such as fleet activity and traffic management hubs for sea vehicles. Independent ships are expected to work in restricted trading areas and parts (Inland Rivers, territorial waters, including surrounding countries). Integrated operation centres will offer increased support and control, with experts moving from ship to shore [16].

For more than a decade, the development of digital, and automation have affected work in the navigational and motor departments of on-board ships, for instance, unmonitored engine areas and maritime "autopilots". The next technological leap would be to reduce the number of crew members on board to provide remote assistance. Experts at shore-based service centres will perform more operations [14].

3.6. User Interfaces

Custom-made services are possible with the help of intelligent interfaces. These interfaces utilize the data mining approaches possibly by means of artificial intelligence. In some cases, robots may be offering the required services to the customers either by digital mode or else physically. The maintenance and repairing jobs on-board vessels may be carried out with the combination of humans and robots. Aerial and subaquatic drones may be employed to examine the parts of the ship that are not reachable are unsafe to reach. The experts may carefully observe the compound engines and other equipment inside the vessel from a distant location (shore). On-request maintenance greatly reduces the transport schedule of the vessel.

4. Overview of Artificial Intelligence

Machine training is a tier of algorithm that enables code algorithms to generate more reliable performance predictions instead of being physically coded. As all marine transports entail smart technology, implementing machine-learning tools can have the greatest benefit in terms of performance, resilience, and cost-cutting. Data is perhaps the major aspect of AI that will help to unlock the inconsistency and accelerate shipping. Currently, data collection on maritime transport is minimal. As a result, deployment of machine learning methods in sea industry is hampered in comparison to other industries. However, the use of machine learning in the maritime field is needed for autonomous shipping or ship automation [14].

Meanwhile, machine learning is a method of computer programming that uses vast quantities of data to boost output criterion [17]. According to the essence of training, it can be divided into 3 main classes. In this scenario, the reader should use the flow diagram below for applications of automated learning. There are 3 types of device training techniques: supervised learning, unsupervised learning, and reinforcement learning. Figure.1 depicts the flow diagram of a machine learning programme.

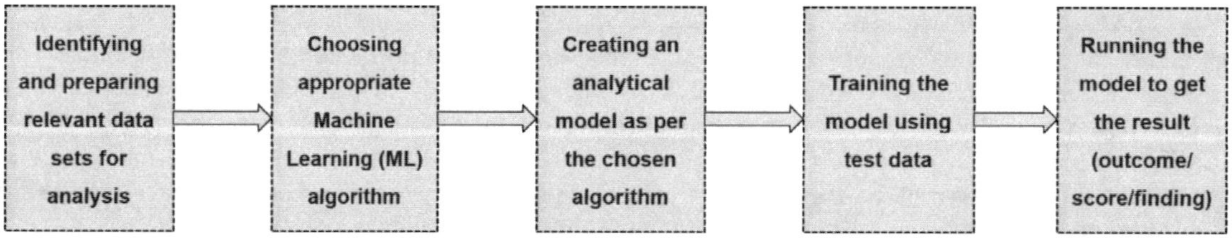

Figure 1. Flow Diagram of a Machine Learning Programme.

4.1. Supervised Learning

Supervised learning involves inputs, and the majority of machine learning approaches involve supervised learnings. The elementary description of methods can indeed be specified as input parameters (x) and target parameters (y) calculating an algorithm to know the function from input to output, as seen in formula below.

$$y = f(x)$$

The meaning is that for a set of input parameters x, the target y can be found. This type can further be divided as (a) regression and (b) classification problems. In (a), the necessary target parameters are continuous. The above learning method is primarily used to achieve point predictions of target variables for given input data. The latter method, on the other hand, necessitates clustering /clumping of target values in a known source database. The widely used methods of supervised learning algorithms include random-forest for classification and regression problems, linear-regressions, support-vector-machines for classification problems, and artificial neural networks for regression problems. Figure.2 displays the fundamental components of Artificial Neural Network (ANN).

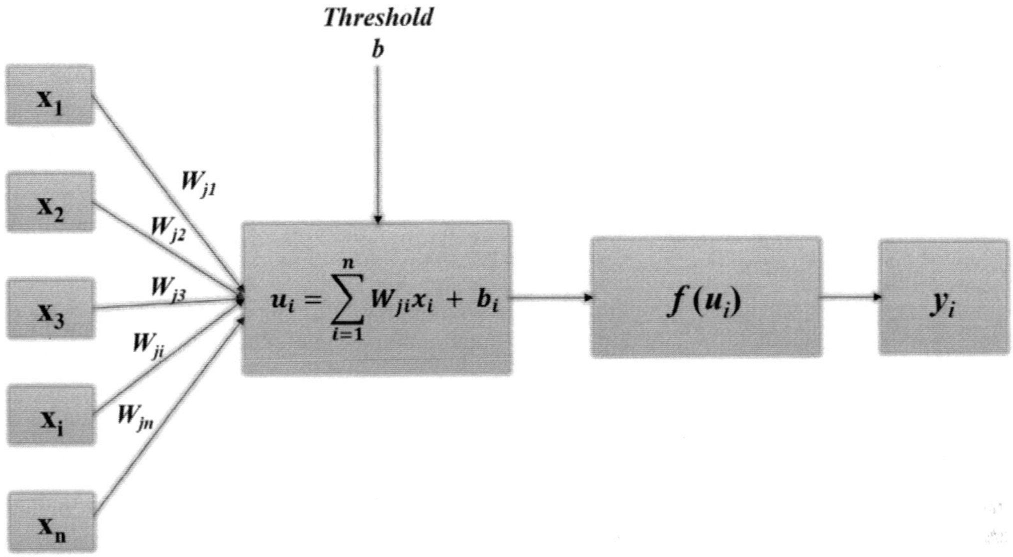

Figure 2. Elements of ANN.

4.2. Unsupervised Learning

This scheme is applicable only when the input data is present. In this situation, none of output variables will be in hand. Unsupervised learning assists in modelling the main structure or allocating the input parameter to boost input learning. In the meanwhile, there is no true response or fix to this sort of problem, because there is no supervision. As a result, it is regarded as unsupervised learning [17]. The machine uses its own algorithm to evaluate and familiarise itself with the original structure in the data in this process. It can also be separated into clustering as well as association problems. The first method can identify inherent classes in data, while the next one is a rule learning problem in which the user is eager to decide distinguishing rules for a larger sample size. The prevalent methods of unsupervised learning rules are k-means for clustering and apriori algorithm for associations rule. One such example for k-clustering is presented in Figure.3.

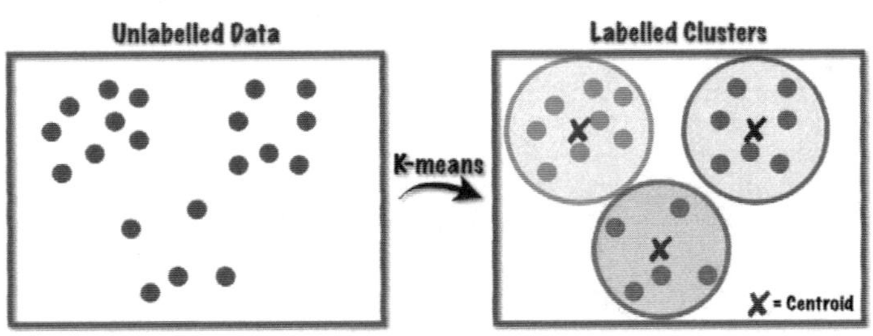

Figure 3. Unsupervised Learning with K-means.

4.3. Reinforcement Learning

This approach is hardly seen in machine learning algorithms since it incorporates known actions to optimise reward in a given situation. While it is similar to supervised learning, there are significant variations, such as the data collection for instruction and the sequence of contingent resolutions. This learning is a subset of artificial intelligence in which an agent chooses which action to take depending on its current state. Q-learning, which employs the bellman equation, and strategy repetition are the two common types [18]. Figure.4 portrays the reinforcement-learning algorithm.

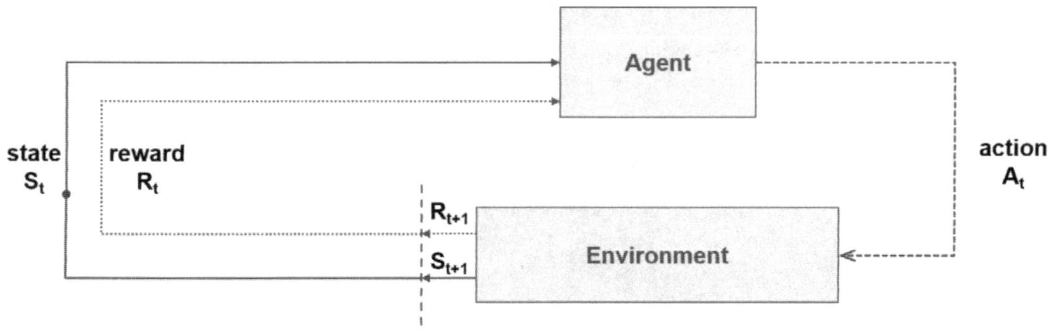

Figure 4. Reinforced-Learning.

5. AI in Maritime Applications

The machine learning application can contribute to enhance shipping because data is one of the chief components in unravelling the uncertainty. It agrees to use savvy algorithms for data analysis and provides a pragmatic approach to maritime transportation problems. The use of such algorithms is profound mostly in the fields of maritime infrastructure scheduling, flight forecasting, ship planning processes, maintenance procedure, etc. The information/particulars of the marine applications facilitate to focus on proper algorithm. The pros and cons of AI in maritime industry is emphasized in Table 3.

Table 3. Pros and Cons of AI - Maritime Sector [15]

Artificial Intelligence – Maritime Sector	
Pros	**Cons**
Cutting-edge Analytics	Reduced data-quality
Computerised Processes/Equipment	Dearth of proper info
Safety/Value-added Security	Panic of occupation replacement
Direction/path optimization	Absence of perfect approach
Performance guessing	Make over into trade is a lengthy course
Cost cutbacks	
Boosting maintenance	

5.1. Autonomous Ships

Any automation starts with the replacement of conventional control by electronic/programmable controllers. Likewise, modern sensors and digital meters exchange the conventional meters. All these circuits are carefully connected through a network. Instead of writing coding that run through pages, a smart algorithm with machine learning technique may be adopted. Independent ships are not predictable ships minus crewmembers; rather, these are mostly a new category of vessel having fewer but highly qualified crewmembers who monitor a variety of organizational activities and services on-board, most likely through direct service stations. Self-directed vessels would not be likely to substitute standard freighters, potentially resulting in work losses for sailors. It is presumed that unmanned ships will be built as component of a constrained transportation infrastructure, creating new nautical routes as a replacement for other modes of transportation. While the daily fleet's "digitalization" accelerates, seafarers' roles will move further toward modern ones, especially in operations testing and system management, and away from functional work. Practicing and preparation will be updated in order to provide seafarers with the requisite latest expertise. Many automated ship techniques can be projected as part of overall shipping, which may possibly eventuate over the near five years. Accordingly, the seafarers may be equipped for data fluency, data interpretation and analysis of huge data. They must be aware of and trained to operate the ships digitally.

5.2. Port Automation

Port operators are normally extra proactive in bringing automation as compared to shipping carriers with respect to the arrangement of autonomous ships. Nonetheless, shipping companies and port operators jointly trust that an integrated gateway may be built in a controlled region in one country, where it could be protected by the nation's national authority. In contrast, when it comes to programmed ships that traverse the waterways of several countries or travel at sea, all national laws and international treaties must be followed. The technology readiness level (TRL) is an indication of hi-tech maturity. Several market reports examine about port automation. In broad-spectrum, prevailing and future practices of computerization in ports will include levels of implementation, and technological maturity. The Artificial intelligence models are much useful in achieving the above.

5.3. Obstacle Prevention for Under-Actuated Unmanned Marine Vessels

The under-actuated unmanned marine vessel having indefinite dynamics of the environs is protected from obstacle using a deep reinforcement-learning algorithm. The foremost gain of the proposed algorithm is its brevity and extendibility; the usual analytic control rule is not essential for manoeuvring the vessel. The vessel takes actions based on the present observations, together with the detected results for obstacles, and the simple measurements of working states of the vessel by means of the proposed algorithm. Further, this architecture is extendible to other composite/intricate tasks [19].

5.4. Liner Shipping Route for Container Ships

Artificial Intelligence principle is useful in planning the best/ideal route to diminish the overall fuel prices. A cross-machine-learning technique that was first developed aids in the prediction of fuel usage for cargo vessels by learning the practical knowledge of an Asian fleet. The predictive results along with certain assumptions of a liner-shipping path initially utilized. It was joined with the asymmetrical travelling salesperson algorithm to create a prime route to curtail the fuel cost. Deep learning is a data-driven practise that is heavily reliant on the accuracy of the learning dataset [20].

5.5. Fault Detection for Ship Systems Operations

Innovative machine learning based expected behaviour modelling with relevant control charts are applied to ship systems for identifying the fault on system operations. The main objective of the study is to permit for anticipatory actions and scheduling, plummeting interruption and to increase the safety, energy competence and add-on routine watching of vessel progressions. The prompt finding of emerging errors will possibly lessen the suboptimal functions of the supervised system, protecting its energy efficient processes [21].

5.6. Ship Fuel Consumption on a Container Ship

Optimizing the fuel consumption of marine vessel will allow for greater productivity and cost-effectiveness in ship management, since fuel price covers majority of running cost. The estimation of vessel fuel usage is a difficult topic since the rate of fuel consumption is directly dependent upon multiple conditions. viz. the state of the primary engine, passenger load, ship draught, sea/ocean weather, meteorological conditions, and so on. Out of an analysis with many predictive models including Multiple-Linear-Regression, Ridge- and LASSO=Regression, Support-Vector-Regression, Tree-Based Algorithms, the Enhancement for a container ship is found to be promising with K-fold cross-validation offering greater accuracy [22]. Another study helped to assess the fuel oil consumption in the main engine using multiple regression algorithms. The following restrictions were taken into account in this study: shipboard information techniques, mid-morning reports, and Automated Data Recording & Tracking schemes [23].

Various other studies on AI schemes include prediction of energy consumption ships in green port [24], Prognostic preservation for ballast pumps on ship-repairing yards [25] and ship collision prevention [26].

While AIS can be used in a variety of shipping applications, the most important application that may yield positive results is vessel location monitoring or detecting a certain vessel's sailing pattern. As a consequence, only the outcome of such an application is elaborated. The same is true for many other applications, depending on the scale of the dataset, the amount of learning and training results, and the data's validity. Table 4 compares vessel pattern tracking with the proposed approach with the other current technologies.

Table 4 Comparison of Vessel Pattern Tracking With Matching Pursuit-Fletcher Reeves (MPFR) [28]

Author	**Title**	**Reference**	**Interpretations**
Chen, Zet.al (2018)	Classification of vessel motion pattern in inland waterways based on Automatic identification System	[28]	**Objective:** Creating reliable techniques for automated vessel movement pattern detection in sea routes. **Model Used:** The Lp-minimization-based sparse representation classification (SRC) tool. **Process Created:** Matching Pursuit - Fletcher Reeves (MPFR) **Datasets Employed:** Wuhan-dataset (W) & Taizhou-dataset (T).

Other Details: The Yangtze River sailing in relation to Wuhan and Taizhou Vessels is classified into five groups based on the pattern of their vessel motions. The relevant datasets are listed below.

Class	**Training Number**		**Test Number**	
	Wuhan (W)	**Taizhou (T)**	**Wuhan (W)**	**Taizhou (T)**
C1	555	355	200	350
C2	675	265	300	230
C3	730	387	260	300
C4	670	517	200	280
C5	830	233	350	290

The obtained **results** are tabulated and compared to the existing approaches, which are mentioned below.

Class	**Suport Vector Machine (%)**		**k-Nearest Neighbour (%)**		**SRC-L1 Norm (%)**		**SRC-L2 Norm (%)**		**SRC-Lp Norm (%)**		**MPFR (%)**	
	W	**T**	**W**	**T**	**W**	**T**	**W**	**T**	**W**	**T**	**W**	**T**
C1	86.5	85.7	88.0	86.2	88.5	88.6	87.0	85.5	89.0	89.1	91.0	91.1
C2	86.7	81.7	88.3	80.8	89.3	83.5	88.7	86.0	90.3	85.2	91.6	88.2
C3	84.6	83.7	85.3	86.0	88.1	87.3	88.8	86.7	89.6	89.7	91.9	90.3
C4	87.0	82.1	87.5	82.9	89.0	85.3	86.0	85.0	88.5	85.3	90.5	86.8
C5	85.4	86.2	86.2	87.2	88.9	89.3	86.6	87.9	89.4	91.3	90.8	92.8
Average	86.0	83.9	87.1	84.6	88.8	86.8	87.4	86.2	89.3	88.1	91.2	89.8

Inference: It can be shown that the precision is determined by the type of dataset as well as the methodology used. On average, the proposed MPFR approach yields slightly higher results for both the Wuhan and Taizhou datasets. This is true with all 5 classes. It also is worth noting that the average training numbers are higher in the Wuhan dataset than the Taizhou dataset, whereas the average test numbers are higher in the Taizhou dataset. It has been discovered that more training/learning fetches more reliable results.

Table 5 compares the results of vessel monitoring patterns produced by the Associative Learning (AL) and Predictive Associative Learning (PAL) methodologies.

Table 5. Comparison of Vessel Pattern Tracing by AL and PAL Methodologies [2] & [3]

Author	**Title**	**Reference**	**Interpretations**
1. Bomberger, N et.al (2006) 2. Rhodes, B at.al (2007)	1. Associative Learning of Vessel Motion Patterns for Maritime Situation Awareness 2. Probabilistic associative learning of vessel motion patterns at multiple spatial scales for maritime situation awareness	[2] [3]	Objectives: Use Associative learning (AL) to recognise the pattern of the Vessl motion [2]. To investigate the motion of a vessel using probabilistic associative learning (PAL) [3].

Other Details: A comparable research has been conducted in both of the preceding literatures by subtly changing the techniques. A standardised square grid is used to discretize vessel position based on vessel velocity over the region of interest surrounding the port of Miami. The zones are split into four westward and eastward zones based on grid positions. The predictions were based on d0 (direct hits/zero distance from predicted cell), d1 (near misses/one cell away from predicted cell), and d2 (two cells away from predicted cell). The output was evaluated using four metrics: recall, coverage, precision and accuracy. The following analyses are tabulated and compared:

Zone Number(s)	**Recall**		**Coverage**		**Precision**		**Accuracy**	
	AL	**PAL**	**AL**	**PAL**	**AL**	**PAL**	**AL**	**PAL**
All	0.20	0.20	0.56	0.55	0.20	0.20	0.36	0.36
1 & 2 (d0,d1 & d2)	0.35	0.31	0.83	0.66	0.23	0.28	0.41	0.47
3 & 4 (d2 only)	0.07	0.08	0.30	0.44	0.13	0.09	0.22	0.17
3 (d2 only)	0.12	0.11	0.44	0.36	0.14	0.19	0.27	0.31
4 (d2 only)	0.03	0.06	0.21	0.47	0.11	0.06	0.15	0.13

Inference: The AL approach uses fixed threshold while the PAL approch uses incremental learning. By lookig over the above Table it is seen that the precision and accuracy are improved in zones 1 & 2 with PAL. The performance is almost same when we consider all zones together with regard to both AL and PAL. Other metrics are varying and depend on the learning rate. A multi-scale approach to express spatial position that fits the spatial scale to the track activity in a given area will result in better prediction results. Spatial forecasts would be unnecessarily high if the spatial scale is too coarse to reflect local activities. More training data would be needed if the spatial scale is too wide to achieve reasonable prediction performance. As the spatial scale is properly configured incremental learning can run quite effectively and soon attain productive levels of efficiency as the system tends to track and evaluate vessel activity.

6. Conclusion

Even if machine learning has previously been included in abundant sections of the online-world, its application in maritime sector is still constrained. To reap the gains for viable conveyance, marine transports need smart software and machine-learning concepts. This paper addressed the different facets of machine learning, as well as their upsides and downsides. The upshot of artificial intelligence on the marine industry's future is also explored. The rigorous investigation reveals that the seafaring experts and investigators ought to focus mainly on apt/suitable algorithm, as the outcomes are different for different algorithms and for different situations. The prospective domains on which the artificial intelligence be applied entails autonomous shipping, port automation and management, route organizing and optimizing, obstacle prevention in underwater environment, fault detection in ship operations, forecasting the fuel usage, energy consumption and management. Rather than focusing on a single application, the approach analysed major and important implementations. In addition, the outcomes of one main application of Vessel path control are elaborated and tabulated.

References

[1] Obradovic, Ines & Milicevic, Mario & Zubrinic, Krunoslav. (2014). Machine learning approaches to maritime anomaly detection. Nase More. 61. 96-101.

[2] Bomberger, Neil & Rhodes, B.J. & Seibert, Michael & Waxman, Allen. (2006). Associative Learning of Vessel Motion Patterns for Maritime Situation Awareness. 1-8. 10.1109/ICIF.2006.301661.

[3] M. Zandipour, B. J. Rhodes and N. A. Bomberger, "Probabilistic prediction of vessel motion at multiple spatial scales for maritime situation awareness," 2008 11th International Conference on Information Fusion, Cologne, 2008, pp. 1-6.

[4] Tu, E., Zhang, G., Rachmawati, L., Rajabally, E., & Huang, G. (2018). Exploiting AIS Data for Intelligent Maritime Navigation: A Comprehensive Survey from Data to Methodology. IEEE Transactions on Intelligent Transportation Systems, 19, 1559-1582.

[5] Dobrkovic, A., Iacob, M.-E., & van Hillegersberg, J. (2015). Using machine learning for unsupervised maritime waypoint discovery from streaming AIS data. Proceedings of the 15th International Conference on Knowledge Technologies and Data-Driven Business - i-KNOW '15. doi:10.1145/2809563.2809573.

[6] Coraddu A, Oneto L, Ghio A, Savio S, Anguita D, Figari M. Machine learning approaches for improving condition-based maintenance of naval propulsion plants. Proceedings of the Institution of Mechanical Engineers, Part M: Journal of Engineering for the Maritime Environment. 2016;230(1):136-153. doi:10.1177/1475090214540874

[7] MacKinnon, Scott & Weber, Reto & Olindersson, Fredrik & Lundh, Monica. (2020). Artificial Intelligence in Maritime Navigation: A Human Factors Perspective. 10.1007/978-3-030-50943-9_54.

[8] Rødseth, Ørnulf & Nordahl, Håvard. (2018). Definition of autonomy levels for merchant ships, Report from NFAS, Norwegian Forum for Autonomous Ships, 2017-08-04. 10.13140/RG.2.2.21069.08163.

[9] WMU: Transport 2040: Automation, Technology, Employment - The Future of Work. World Maritime University, Malmö, Sweden (2019)

[10] Autor, D. H., Levy, F. and Murnane, R. J., 2003. The Skill Content of Recent Technological

Change: An Empirical Exploration, The Quarterly Journal of Economics, 118(4), pp. 1279–1333. doi: 10.1162/003355303322552801.

[11] Bekiaris, E. and Loukea, M., 2017. Future scenarios on skills and competences required by the Transport sector in the short, mid and long-term. D2.1. SKILLFUL Project.

[12] Cyber-enabled ships, ShipRight procedure – autonomous ships. First edition, July 2016, A Lloyd's Register guidance document.

[13] Benjamin, M., Curcio, J.A.: COLREGS-based navigation of autonomous marine vehicles.In: Proceedings of the IEEE/OES Autonomous Underwater Vehicles (2004).

[14] Marine-digital.com. 2021. Artificial Intelligence (AI) And Machine Learning (ML) In Maritime Logistics. [online] Available at: <https://marine-digital.com/article_ai_and_ml> [Accessed 11April 2021].

[15] James R. DeChant, V., 2021. How AI Is Influencing The Shipping Industry Today. [online] Adv-polymer.com. Available at: <https://www.adv-polymer.com/blog/artificial-intelligence-in-shipping#ch1> [Accessed 11 April 2021].

[16] F.R., M., 2021. How AI & Automation Has Overhauled The Shipping Industry. [online] Analytics India Magazine. Available at: <https://analyticsindiamag.com/ai-shipping-autonomous-drive/> [Accessed 11 April 2021].

[17] Alpaydin, Ethem. (2009). "Introduction to machine learning". MIT press.

[18] Akyuz, Emre & Cicek, Kadir & Celik, Metin. (2019). A Comparative Research of Machine Learning Impact to Future of Maritime Transportation. Procedia Computer Science. 158. 275-280. 10.1016/j.procs.2019.09.052.

[19] Cheng, Y., & Zhang, W. (2018). Concise deep reinforcement learning obstacle avoidance for underactuated unmanned marine vessels. Neurocomputing, 272, 63–73. doi:10.1016/j.neucom.2017.06.066.

[20] Bui-Duy, L., & Vu-Thi-Minh, N. (2020). Utilization of a deep learning-based fuel consumption model in choosing a liner shipping route for container ships in Asia. The Asian Journal of Shipping and Logistics. doi:10.1016/j.ajsl.2020.04.003.

[21] Cheliotis, M., Lazakis, I., & Theotokatos, G. (2020). Machine learning and data-driven fault detection for ship systems operations. Ocean Engineering, 216, 107968. doi:10.1016/j.oceaneng.2020.107968.

[22] Uyanık, Tayfun & Karatuğ, Çağlar & Arslanoğlu, Yasin. (2020). Machine learning approach to ship fuel consumption: A case of container vessel. Transportation Research Part D Transport and Environment. 84. 10.1016/j.trd.2020.102389.

[23] Gkerekos, C., Lazakis, I., & Theotokatos, G. (2019). Machine learning models for predicting ship main engine Fuel Oil Consumption: A comparative study. Ocean Engineering, 188, 106282. doi:10.1016/j.oceaneng.2019.106282.

[24] Peng, Y., Liu, H., Li, X., Huang, J., & Wang, W. (2020). Machine learning method for energy consumption prediction of ships in port considering green ports. Journal of Cleaner Production, 121564. doi:10.1016/j.jclepro.2020.121564.

[25] Kimera, D., & Nangolo, F. N. (2020). Predictive maintenance for ballast pumps on ship repair yards via machine learning. Transportation Engineering, 100020. doi:10.1016/j.treng.2020.100020.

[26] Gao, M., & Shi, G.-Y. (2020). Ship collision avoidance anthropomorphic decision-making for structured learning based on AIS with Seq-CGAN. Ocean Engineering, 217, 107922. doi:10.1016/j.oceaneng.2020.107922.

[27] A. Prasanth (2020). Certain Investigations on Energy-Efficient Fault Detection and Recovery Management in Underwater Wireless Sensor Networks, Journal of Circuits, Systems and Computers, doi: 10.1142/S0218126621501371.

[28] Chen, Z., Xue, J., Wu, C., Qin, L., Liu, L., & Cheng, X. (2018). Classification of vessel motion pattern in inland waterways based on Automatic Identification System. Ocean Engineering, 161, 69–76. doi:10.1016/j.oceaneng.2018.04.072.

IOP Conf. Series: Materials Science and Engineering 1177 (2021) 012003 doi:10.1088/1757-899X/1177/1/012003

Modified Bridgeless Landsman Converter Fed EV Battery Charger with Improved Power Factor

M.Dinesh[1] and K.B. Bhaskar[2]
[1]Assistant Professor – Electrical and Electronics Engineering, Adhi College of Engineering and Technology, Chennai
[2]Associate Professor - Electrical and Electronics Engineering, Adhi College of Engineering and Technology, Chennai

E-mail: dinesh.eee@adhi.edu.in, eee.hod@adhi.edu.in,
Abstract. This research work focuses on the design and construction of a novel Battery-powered Electric Vehicle (BpEV) and to enhance the power-factor of the input AC. This is accomplished by substituting the prevalent diode converter with a newer, high-quality Landsman converter that is equipped with power-factor tuning. The Flyback-isolated and PFC converters work together to let the EV battery to switch between constant current and constant voltage modes. The recommended PFC converter is regulated using a single entity to attain consistent DC-link voltage and to boost the power-factor close to unity, with simulation carried out using MATLAB. As per the findings, a minimum of 60V DC output is required for charging the battery and suppressing input harmonics. In comparison to the usual architecture, the suggested scheme could provide increased power quality, very little device stress, less ripple at both input and output, and significantly lower harmonics. To validate the simulation results, a prototype is developed and deployed to charge a 48V EV battery with a rating of 100Ah in order to make it compliant with the IEC 61000-3-2 standard, particularly in the case of high voltage spikes at the input. The charger's performance has been deemed to be good in all conditions based on matlab simulation results.

Keywords: Power Factor, Landsman Converter, Flyback converter, Isolation transformer, Harmonics.

1. Introduction

The charging process of any battery used in an electric locomotive is primarily determined by its dimensions and category, such as the magnitude and duration of voltage/current to be applied, the procedure after full-charging is complete, and so on. A few battery types are much more adaptable and capable of sustained charging even after full charge. They are typically recharged using either a constant current or a constant voltage source, depending on the situation. Few other batteries mandate/employ a timer to turn off the charging current at a designated time, usually once the charge is full, and the chargers must be separated manually at almost every charging cycle. Overcharging, damage (reduced capacity, shorter lifespan), overheating, and even exploding are all risks associated with other battery types. The charger comprises temperature and voltage sensor circuits, as well as a microprocessor, to securely fine-tune the charging current/voltage and diagnose the cut-off at the end of the charge. A slow battery charger may take several hours to reach full charge. High-capacity chargers can promptly rebuild the majority of a battery's capacity; however, they may be too powerful for some battery types. To safeguard the battery from overcharging, such batteries require active monitoring. High-rate chargers are appropriate for electric vehicles. Setting up those chargers, as well as providing the necessary assistance, is a concern, particularly for public accessing electric locomotives. A competent battery charger is the foundation for long-lasting and reliable batteries. Chargers are frequently given minimal importance and are seen as an "afterthought" in a price-sensitive market. The battery and charger, like a horse and carriage, must function together. The power supply is prioritized through prudent planning and is critical at the initial phase, well before implementation. Technologists are more often clueless of the complexities of the power source, particularly while charging in complicated environments.

A trickle-charger delivers a limited electric current to a battery that has been sitting for a long time, just enough to keep it from self-discharging. Trickle charging is not compatible with all battery types, and attempting to do so may result in harm. Indefinite trickle charging is not possible due to the chemical technique utilized in lithium ion battery cells.

2. Literature Review

With stringent regulations on air pollution, fuel cost/consumption, climate change concerns, and sparse energy resources, electric mobility makes a significant contribution to the development of sustainable and efficient transportation alternatives [1-2]. And [3] provides a study based on the actual predicament and technological advances for electric vehicle (EV) propulsion in this regard. Transportation by electricity has a number of benefits than the orthodox gasoline and petrol automobiles. Yet, a researcher's undivided attention is required to fully incorporate transportation electrification. To fit them into the prevailing distribution scheme, certain effective control mechanisms must be created. Some of the aforementioned strategies are linked to the concerns with power quality that EV chargers solve throughout the charging procedure [4-5]. The EVs are outfitted with rechargeable batteries to achieve the desired adhesion/friction strength. EV chargers, that are otherwise AC–DC converters, are typically used to recharge all such batteries. The much more widespread EV charging layout includes a boost converter that precedes an isolated converter [6]. The manufacturer determines the capabilities of such a category of charger exclusively depending on the quality of the DC-DC converter owing to tightly controlled voltage as well as current outputs. [7-10] describe more about interleaved and zero-voltage switching PFC converters with respect to charging units that diminish both the inductor dimension and ripple at the output. The cost of interleaving a PFC converter is substantial. The bridge configuration is perhaps popular for PFC-type EV charging units, and it offers features such as improved power density and quality though the setup of four switches exacerbates charging mechanism [11]. A resonant converter (L-L-C type) [12-13] is a captivating alternative because of its maximum effectiveness, little Electromagnetic interference, noise, and extraordinary power density for a variety of inputs. Due to the additional hardships in the modelling and development of L-L-C converters, this sort of configuration is switched by uni- or bi-directional AC-DC converters in embedded on-board or off-board modules [14]. The conversion from AC to DC is imperative for a large number of single-directional isolated 1-stage or 2-stage converters with no isolation supplied via diode-bridge-rectifiers for charging electrically operated rickshaw [15]. The prescribed rickshaw is observed with a battery as a load to illustrate the power quality of the charger supplied via a standard diode-bridge converter. The result clearly shows that the power quality is not as specified by standard IEC 61000-3-2 [16]. Because of the addition of a full-bridge at the input side of the charger, there is a major reduction in harmonic levels at the input, notably throughout charging, but this effect in a low power-factor at the source.

Furthermore, the input current is no longer sinusoidal, resulting in an increase in source voltage and current displacement. As a result, at the front-end of a traditional DBR fed charger, an effective power factor correction (PFC) technique is required, which also eliminates the negative impacts of input DBR. Furthermore, the input current becomes non-sinusoidal, necessitating the use of an appropriate power-factor correcting mechanism in conjunction with the existing DBR in order to negate the adverse influence of the input DBR. This, in turn, reduces system/conduction losses, enhancing efficiency and abolishing the need for such a bridge. A fair comparison of bridge and bridgeless charges with and without power-factor correcting devices are available in literature in [17-21]. The Bridgeless configurations possess large electromagnetic interference, increased noise and it is difficult to control them while it offers higher efficiency and lower thermal stress when the power level is high. Besides that, there is a limit to adjusting the duty ratio for regulating DC-link voltages, particularly at low source voltage values. As a result, many buck-boost converters are debated in order to attain a substantial increase in duty cycle [22-28]. Due to the total absence of a bridge, they can achieve high charger efficiencies and minimize losses even when the voltage at the source is low. The Landsman converter produces superior results, which have been authored in the literature [29].

The proposed work makes an effort to combine the beneficial aspects of many converters presented in the literature while minimising their drawbacks. It is attempted to develop a fly-back converter without the bridge structure and to outfit it with a modified Landsman converter for power-factor adjustment. The aforementioned converter is simulated and demonstrated to charge a 48 V, 100Ah capacity battery in accordance with the IEC 6100-3-2 power quality standards.

3. Operation of Modified BL Converter Fed Charger

The designed charger has two stages of operation: a modified bridgeless converter with superior wave-shaping of input and an isolated converter for charging the electrified locomotive through constant-current or voltage mode based on their monetary or voltage stress constraints. Because the cost of the battery charger is more crucial across most cases, the discontinuous current operation is chosen (DCM i.e. Constant Voltage). Another key reason is that it requires fewer sensors to stabilize battery charging.

3.1 Proposed method

Figure 1 depicts an Electrified Vehicle charger fed by a bridgeless converter, the voltage of which is regulated using an intermediate stage DC-link voltage. This set-up employs a single-phase AC voltage at the input of the charger. The input side diode bridge is eliminated when using two Landsman converters, one for each of the positive and negative half lines. Because fewer elements are used per switching cycle, losses during the conduction period are nearly reduced to 50%. To achieve better switching performance, the two converters use synchronised switching at a frequency of around 20 kc/s. The output inductor currents through Lop and Lon are made discontinuous at each half cycle for a period of one switching cycle, allowing a single sensor to detect the intermediate output voltage. The intermediate DC-voltage at the output of the charger is very well modulated over a wide input range by a two-term (PI) controller in voltage mode. This tightly controlled voltage, tuned with a bridgeless

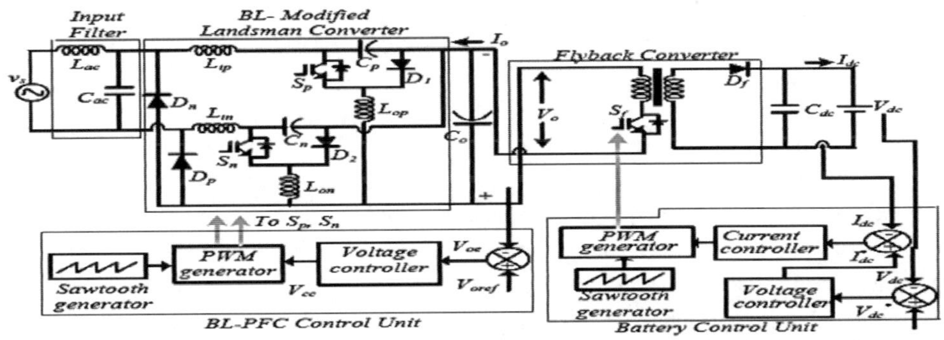

power-factor correcting device, serves as the isolated converter's input. The EV battery thus earns a synchronised charging current proportional to its state of charge.

Figure 1. Modified- BL Landsman PFC Converter-Fed Battery- Charger for Electrified Automobile - Block Diagram

The block diagram of modified-bridgeless landsman-converter fed battery charger for electric vehicle input of the Alternating source supply of 230V AC source and the input voltage is present with some harmonics in order to reduce the harmonics present in the AC source filter circuit are used in the front end side. Modified circuit of landsman converter is implemented and the converter is controlled by the driver circuit of TLP 250 and the controller is PWM controller is used for the gate pulse. In between the landsman converter and flyback converter isolation transformer is implemented the output of the flyback converter dc output voltage of around 60V is fed to the battery for charging. In this modified BL landsman power factor converter the power factor value is increased to 0.9443. The ultimate purpose is to boost the power factor and mitigate the harmonics in the AC

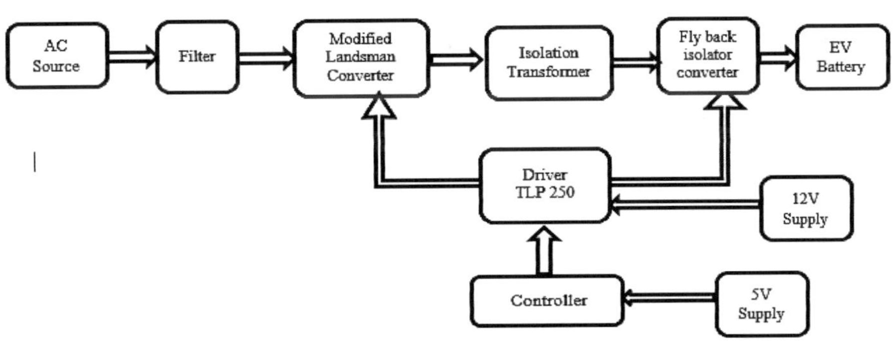

supply side of the circuit.

Figure 2. Modified-BL Landsman PFC Converter-Fed Battery Charger for Electrified Automobile

3.2 Proposed Isolated Converter Operation

Because of its simple form, the flyback converter is widely used to implement battery chargers for low-cost battery-operated vehicles such as E-rickshaws. In mode-I, Sf is turned "ON," and the magnetic inductance goes up to store energy in the high-frequency transformer with ferrite core, causing the current to grow. The polarity of the flyback transformer and the position of the output diode in the network clearly show that no energy is transferred to the battery side. The output DC-link capacitor supplies the battery with the essential charging current.

IOP Conf. Series: Materials Science and Engineering 1177 (2021) 012003 doi:10.1088/1757-899X/1177/1/012003

In mode II, the voltage polarity of the primary and secondary windings is reversed while Sf is turned "OFF." The energy deposited in the transformer's core is thus released to the battery via the diode rectifier at the output.

In mode-III, the energy stacked in the transformer core is totally receded at the end of every switching cycle. The DC-link capacitor at the output delivers the charging energy to the battery. In the next switching cycle, Sf is switched “ON” again in the same sequence. To keep the final output DC voltage above the nominal battery voltage, the flyback converter with cascaded dual PI controller should be functioned in DCM conduction mode. In this way, the battery is charged in CC mode at a constant current rate until it reaches 80 percent SOC, and now in CV mode at a lower current rate until it reaches 100 percent SOC. Finally, an appropriate PI controller with voltage and current tuning is added to ensure for secure charging of the battery with limited ripple. The input voltage of the envisaged improved BL front-end converter is investigated throughout rated and dynamic EV battery charger operation.

Table 1. Performance Assessment of Recommended BL-Converter Fed-Charger and Orthodox Chargers

Configuration Features	Orthodox Charger (No PFC)	Orthodox Charger (DBR+PFC)	Recommended BL-Converter Fed-Charger
Number of elements	_	More	Increased
Control (with PFC)	-	Voltage Follower	Voltage Follower
Control(Battery)	Hard	Simple (dual PI)	Simple (dual PI)
Sensor (with PFC)	-	1 for output voltage	1 for output voltage
Losses (with DBR and PFC)	-	6.5% of total power	5.88% of total power
PFC	-	Yes	Yes

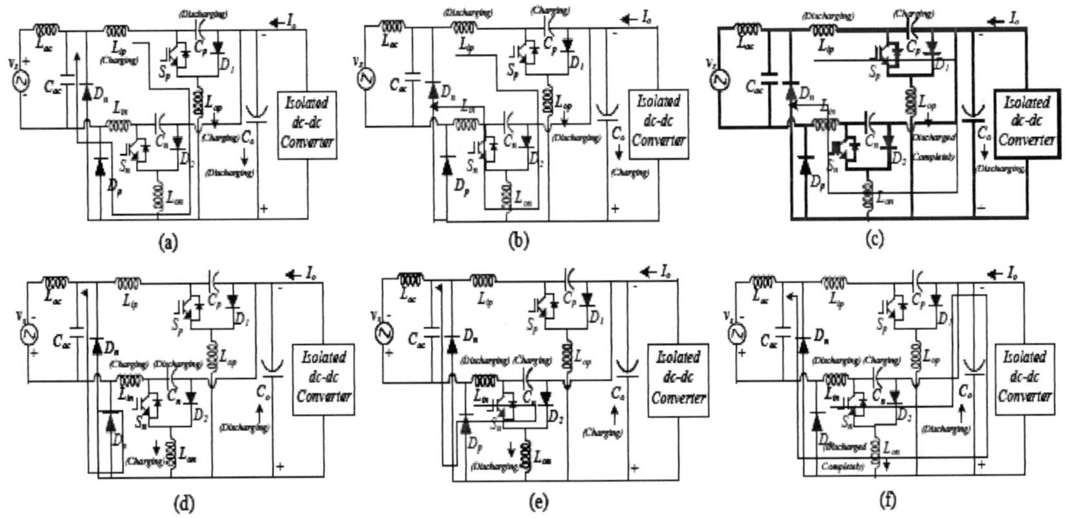

Figure 3. Working Mechanism of Modified- BL Landsman converter (a)-(c) Operation in positive-half cycle: Mode P-I, P-II, and P-III (d)-(f) Operation in negative-half cycle: Mode N-I, N-II, and N-III

4. Simulation Model

In the Simulation model the first part of the circuit is AC source voltage of 230V with harmonics present in the circuit. In order to reduce the harmonics present in the AC source filter circuit are used in the front-end side. Main concept of landsman converter is to boost the power factor and to diminish the harmonics present in the supply side. The inductor values 1e-3, and 22e-3 and capacitor values 15e-3, 220e-6 are used in landsman converter with two MOSFET switches which is triggered by the gate pulse signal and which is separated by an isolation transformer is implemented. The flyback converter dc output voltage of around 60V is fed to the battery for charging. In this modified BL landsman power factor converter the power factor value is increased to 0.9443.

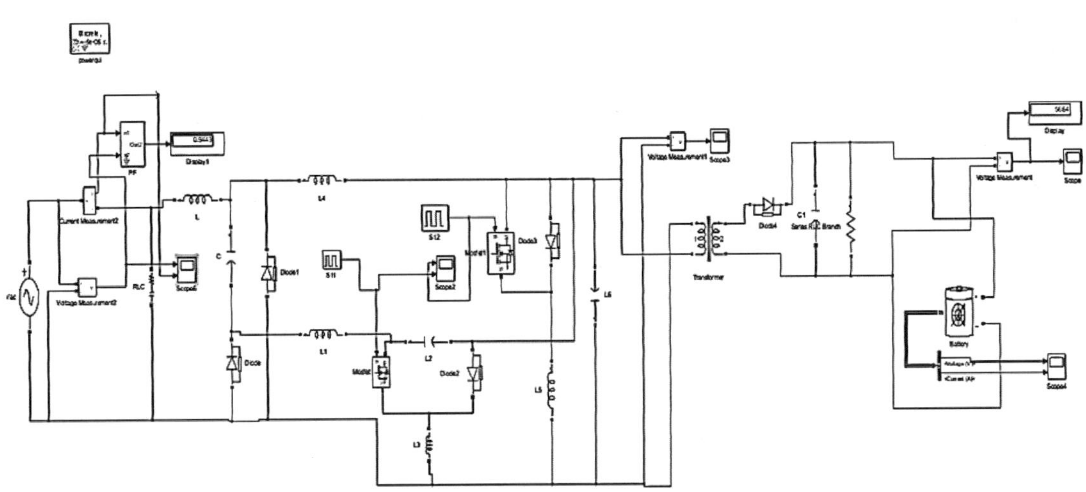

Figure 4. Simulation Model

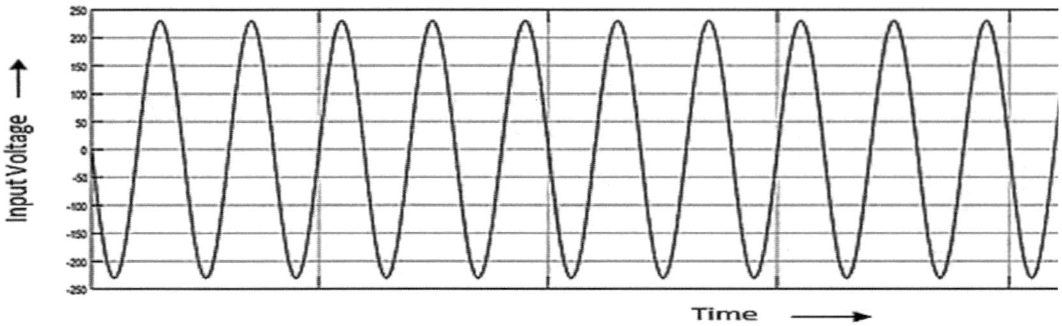

Figure 5. AC Input voltage

In the figure 5 shows the input source voltage of 230V AC source which the harmonics and ripple factor present in the circuit in order to reduce the input side harmonics in the AC source filter circuit with inductor and capacitor are connected across the circuit.

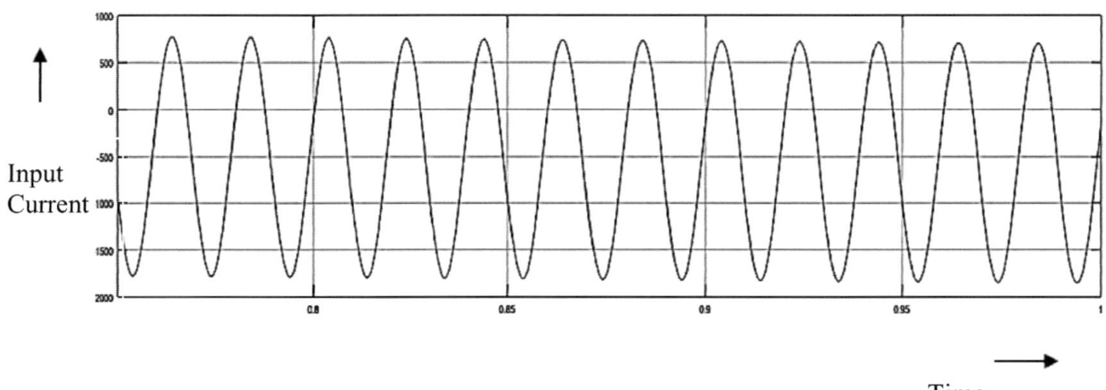

Figure 6. AC Input Current

In figure 6 waveform represents the input current of the AC input side the current waveform oscillate from 0 to 500ma peak to peak current with respect to time duration 1ms value. The modified BL landsman converter is used to increase the power factor value in the circuit with increase in value of 0.944 when compared with older type of converter which is suitable for increasing the power factor value.

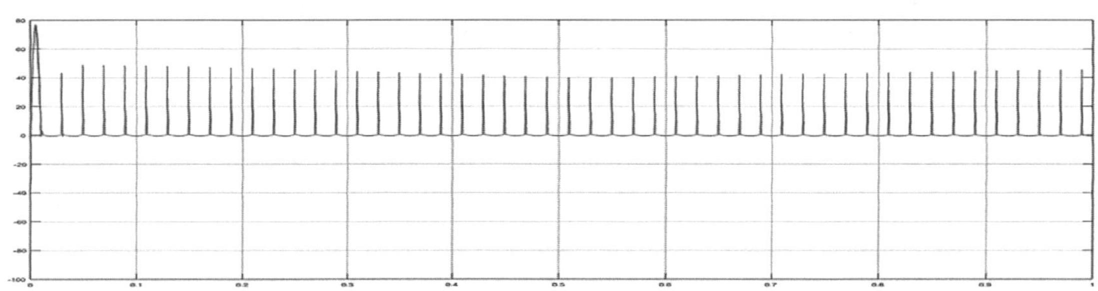

Figure 7. Output waveform for Modified BL Landsman Converter

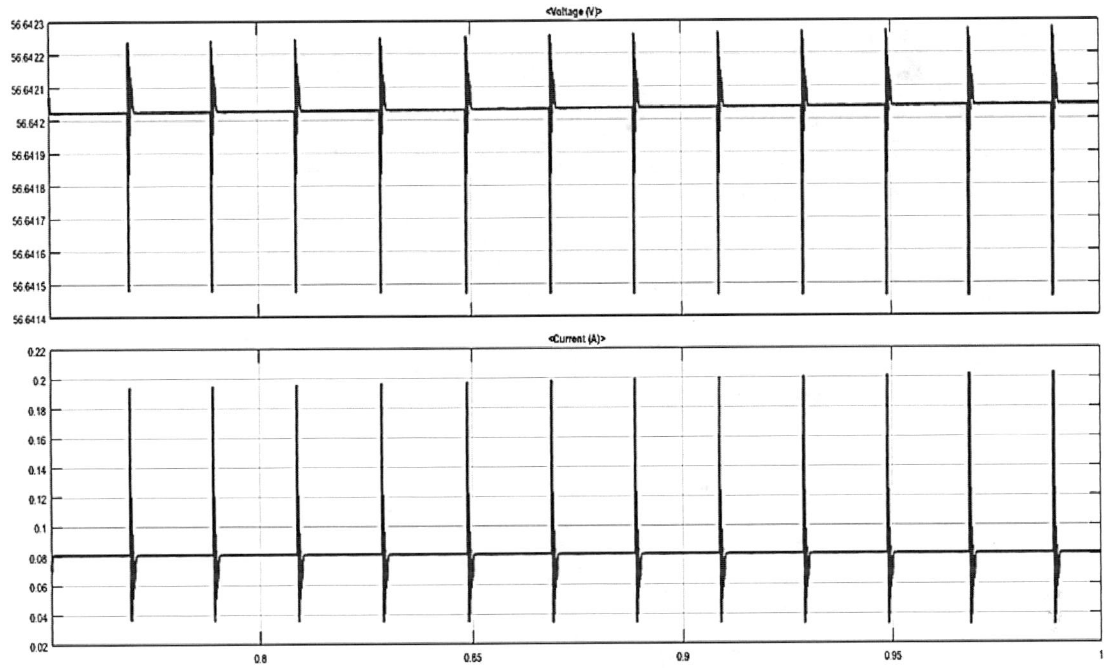

Figure 8. DC output Voltage and Current waveforms

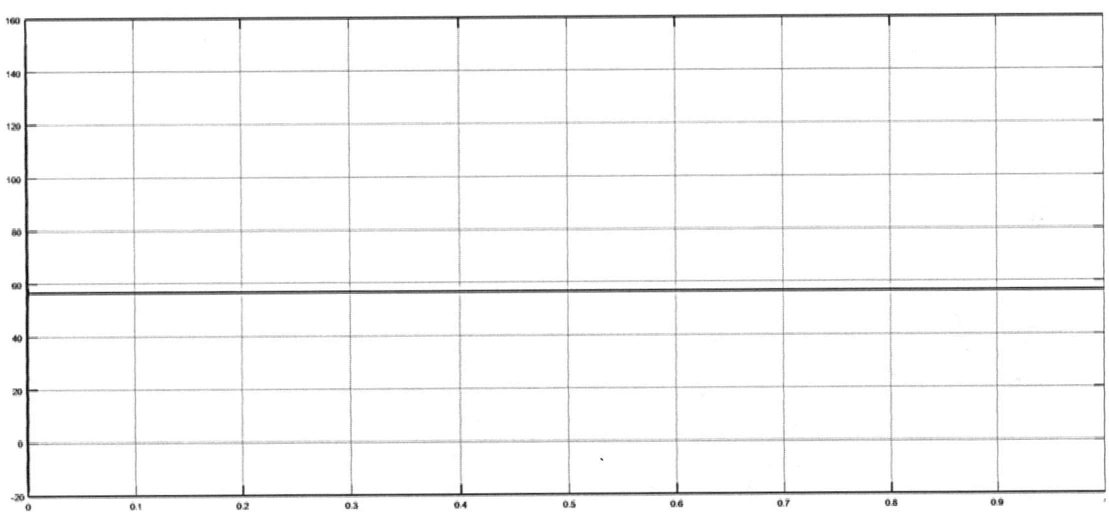

Figure 9. Battery output Voltage waveforms

In figure 8 waveform denotes the voltage and current outputs of the battery; voltage reading is around 56.64V and current value 0.08mA to 0.1mA. In figure 9 represent the battery output voltage and output dc voltage of fly back converter in matlab Simulink result the output voltage is around 56V dc which is used to charge the EV battery.

IOP Conf. Series: Materials Science and Engineering 1177 (2021) 012003 doi:10.1088/1757-899X/1177/1/012003

5. Hardware

Figure 10 Power Factor value measurement (PF = 0.944)

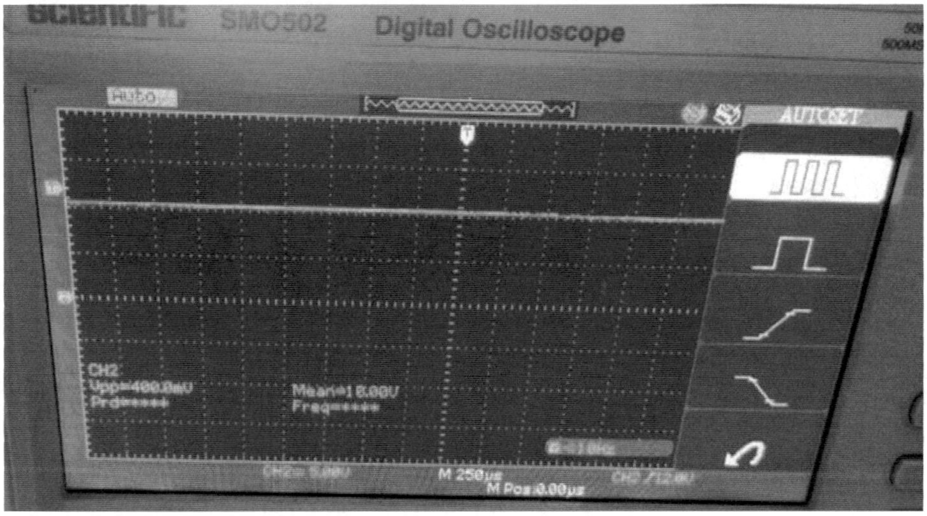

Figure 11 Output DC voltage measurement in Digital Oscilloscope

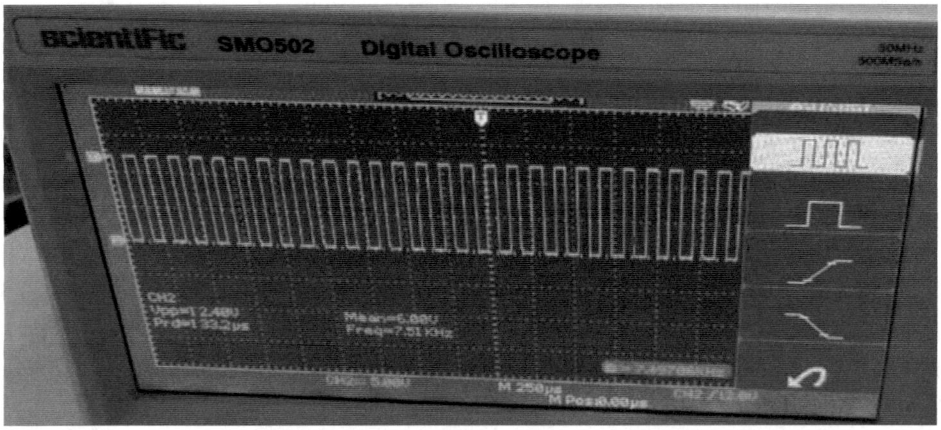

Figure 12 MOSFET Gate pulse Trigger signal measurement

IOP Conf. Series: Materials Science and Engineering 1177 (2021) 012003 doi:10.1088/1757-899X/1177/1/012003

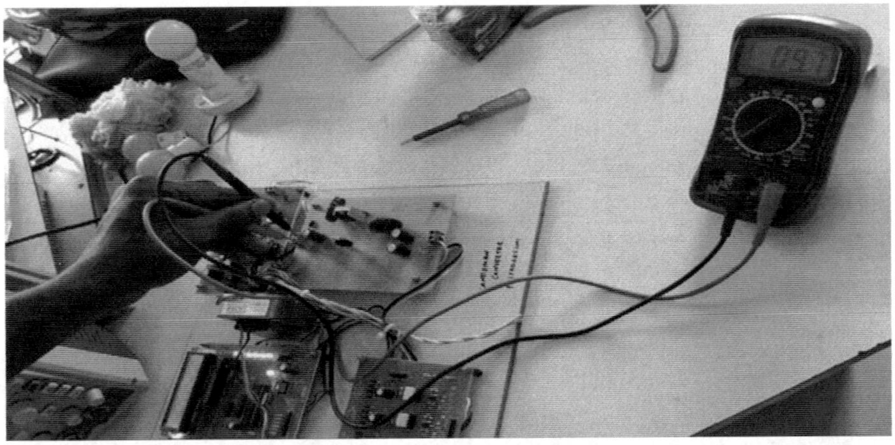

Figure 13 Output voltage of Modified BL Landsman Converter output voltage of 9.7V AC output.

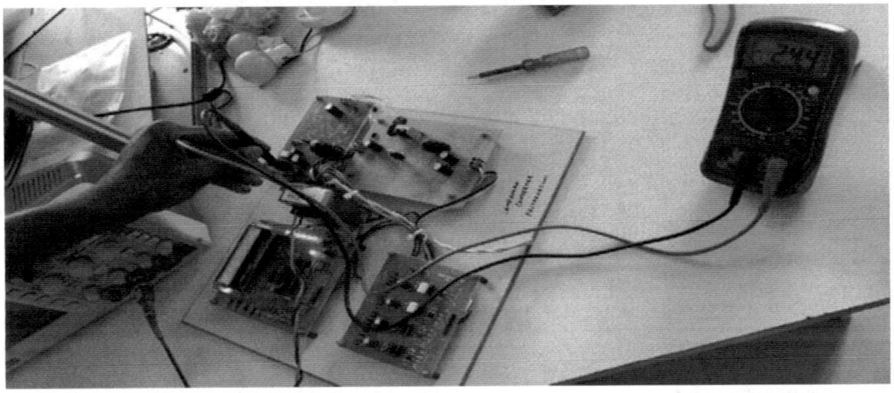

Figure 14 Output voltage of Fly back Converter output voltage of 24.4V DC output.

6. Conclusion

This article suggests, analyses, and validates an enhanced EV charger that uses a modified-BL Landsman converter accompanied by a flyback converter to charge EV batteries with innate PF Adjustment. The suggested EV charger's design and operation in DCM mode has the benefit of less quantity of sensors. Furthermore, the suggested BL converter had also shrunk the ripples due to the presence of inductors at supply/load sides. The supremacy of the suggested charger is validated for both constant and rapidly changing input voltages. The hardware experimental findings also ensure that the new method of EV charging meets the required levels of power-quality. In simulation result it is simulated with 48V battery with input 230V AC source with power factor improvement of 0.944 value and output of the Landsman converter is 40V AC output in simulation and fly back converter output value is 56.4V which is used to charge the EV battery in matlab result. In the hardware result are validated with landsman converter output voltage is 9.7 V AC and the output of Flyback converter is 24.4V DC voltage which is used to charge a 12V battery. Furthermore, the input current THD is decreased to as low as 4.3 percent to fulfil the IEC 61000-3-2 power quality standard criteria. As a result, the suggested BL converter fed charger seeks to be a cost-effective, dependable, and appropriate alternative to the traditional lossy and inefficient EV battery charger.

References

[1] Wencong Su, Habiballah Eichi, Wente Zeng and Mo-Yuen Chow, "A survey on the electrification of transportation in a smart grid environment," IEEE Transactions Industrial Informatics, vol. 8, no. 1, pp. 1-10, Feb. 2012.

[2] Ching Chuen Chan, "The state of the art of electric, hybrid, and fuel cell vehicles," Proc. IEEE, vol. 95, no. 4, pp. 704–718, Apr. 2007.

[3] Kaushik Rajashekara, "Present status and future trends in electric vehicle propulsion technologies," IEEE J. Emerg. Sel. Topics Power Electronics., vol. 1, no. 1, pp. 3–10, Mar. 2013.

[4] Juan C. Gomez and Medhat M. Morcos, "Impact of EV battery chargers on the power quality of distribution systems," IEEE Transactions Power Del., vol. 18, no. 3, pp. 975–981, Jul. 2003.

[5] Luca Solero, "Nonconventional on-board charger for electric vehicle propulsion batteries," IEEE Transactions Vehicular Technology, vol. 50, no. 1, pp. 144-149, Jan 2001.

[6] Devendra Patil and Vivek Agarwal, "Compact onboard singlephase EV battery charger with novel low-frequency ripple compensator and optimum filter design," IEEE Transactions Vehicular Technology, vol. 65, no. 4, pp. 1948-1956, April 2016..

[7] Hye-Jin Kim, Gab-Su Seo, Bo-Hyung Cho, and Hangseok Choi, "A simple average current control with on-time doubler for multiphase CCM PFC converter," IEEE Trans. Power Electron., vol. 30, no. 3, pp. 1683–1693, Mar. 2015.

[8] Adria Marcos-Pastor, Enric Vidal-Idiarte, Angel Cid-Pastor and L. Martinez-Salamero, "Interleaved digital power factor correction based on the sliding-mode approach," IEEE Transactions Power Electronics, vol. 31, no. 6, pp. 4641-4653, June 2016.

[9] Muntasir Alam, Wilson Eberle, Deepak S. Gautam and Chris Botting, "A soft-switching bridgeless AC–DC power factor correction converter," IEEE Transactions Power Electronics, vol. 32, no. 10, pp. 7716-7726, Oct. 2017..

[10] Siddharth Kulasekaran and Raja Ayyanar, "A 500-kHz, 3.3-kW power factor correction circuit with low-loss auxiliary ZVT circuit," IEEE Transactions Power Electronics, vol. 33, no. 6, pp. 4783-4795, June 2018.

[11] Jae-Hyun Kim, Chong-Eun Kim, Jae-Kuk Kim, Jae-Bum Lee, and Gun-Woo Moon, "Analysis on load-adaptive phase-shift control for high efficiency full-bridge LLC resonant converter under lightloadconditions," IEEE Trans. Power Electron., vol. 31, no. 7, pp. 4942–4955, Jul. 2016.

[12] Chuan Shi, Haoyu Wang, Serkan Dusmez and Alireza Khaligh, "A SiC-based high-efficiency isolated onboard PEV charger with ultrawide DC-link voltage range," IEEE Transactions Industry Applications, vol. 53, no. 1, pp. 501-511, Jan.-Feb. 2017.

[13] Junjun Deng, Siqi Li, Sideng Hu, Chunting C. Mi and Ruiqing Ma, "Design methodology of LLC resonant converters for electric vehicle battery chargers," IEEE Transactions Vehicular Technology, vol. 63, no. 4, pp. 1581-1592, May 2014.

[14] Chuan Shi, Yichao Tang and Alireza Khaligh, "A single-phase integrated onboard battery charger using propulsion system for plug-in electric vehicles," IEEE Transactions Vehicular Technology, vol. 66, no. 12, pp. 10899-10910, Dec. 2017.

[15] Bhim Singh, Sanjeev Singh, Ambarish Chandra and Kamal Al- Haddad, "Comprehensive study of single-phase AC-DC power factor corrected converters with high-frequency isolation," IEEE Transactions Industrial Informatics, vol. 7, no. 4, pp. 540-556, Nov. 2011

IOP Conf. Series: Materials Science and Engineering 1177 (2021) 012003 doi:10.1088/1757-899X/1177/1/012003

[16] Limits for Harmonics Current Emissions (Equipment current □16A per Phase), International standards IEC 61000-3-2, 2000.
[17] Alessandro Malschitzky, Felipe Albuquerque, Eloi Agostini and Claudinor B. Nascimento, "Single-stage integrated bridgelessboost nonresonant half-bridge converter for led driver applications," IEEE Transactions Industrial Electronics, vol. 65, no. 5, pp. 3866-3878, May 2018.
[18] Yeonho Jeong, Jae K. Kim and Gun W. Moon, "A bridgeless dual boost rectifier with soft-switching capability and minimized additional conduction loss," IEEE Transactions Industrial Electronics, vol. 65, no. 3, pp. 2226-2233, March 2018..
[19] Muntasir Alam, Wilson Eberle, Deepak S. Gautam, Chris Botting, Nicholas Dohmeier and Fariborz Musavi, "A hybrid resonant pulse-width modulation bridgeless AC–DC power factor correction converter," IEEE Transactions Industry Applications, vol. 53, no. 2, pp. 1406-1415, March-April 2017.
[20] Khairul S. B. Muhammad and Dylan D. C. Lu, "ZCS bridgeless boost PFC rectifier using only two active switches," IEEE Transactions Industrial Electronics, vol. 62, no. 5, pp. 2795-2806, May 2015.
[21] Julio C. Dias and Telles B. Lazzarin, "A family of voltagemultiplier unidirectional single-phase hybrid boost PFC rectifiers,"IEEE Transactions Industrial Electronics, vol. 65, no. 1, pp. 232-241, Jan. 2018.
[22] Ben Zhao, Alexander Abramovitz and Keyue Smedley, "Family of bridgeless buck-boost PFC rectifiers," IEEE Transactions Power Electronics, vol. 30, no. 12, pp. 6524-6527, Dec. 2015.
[23] Vashist Bist and Bhim Singh, "An adjustable-speed PFC bridgeless buck–boost converter-fed BLDC motor drive," IEEE Transactions Industrial Electronics, vol. 61, no. 6, pp. 2665-2677, June 2014.
[24] Ahmed M. Al-Gabri, Abbas A. Fardoun, and Esam H. Ismail, "Bridgeless PFC modified SEPIC rectifier with extended gain for universal input voltage applications," IEEE Transactions PowerElectronics, vol. 30, no. 8, pp. 4272–4282, Aug. 2015.
[25] Bhim Singh and Vashist Bist, "A BL-CSC Converter-Fed BLDC Motor Drive With Power Factor Correction," IEEE Transactions Industrial Electronics, vol. 62, no. 1, pp. 172-183, Jan. 2015.
[26] Hong-Tzer Yang, Hsin-Wei Chiang and Chung-Yu Chen, "Implementation of bridgeless cuk power factor corrector with positive output voltage," IEEE Transactions Industry Applications, vol. 51, no. 4, pp. 3325-3333, July-Aug. 2015.
[27] Anjanee K. Mishra and Bhim Singh, "Control of SRM drive for photovoltaic powered water pumping system," IET Electric Power Applications, vol. 11, no. 6, pp. 1055-1066, 7 2017.
[28] Preethy V. Warrier and Preetha P. K., "Power factor correction circuit with one cycle controlled bridgeless zeta converter for display monitors in hospitals," IEEE TENCON'15, pp. 1-6
[29] Aman Jha, Vashist Bist and Bhim Singh, "Landsman based PFC with PWM dimming for high brightness LED driver," IEEE INDICON'15, pp. 1-6.

IOP Conf. Series: Materials Science and Engineering 1177 (2021) 012004 doi:10.1088/1757-899X/1177/1/012004

Review and Comparison of Various Types of Generation Using WECS Topologies

A Vijayaprabhu[1] K B Bhaskar[2] D Jasmine susila[3] M Dinesh[4]
[1]Assistant professor Department of Electronics and Communication, Sri Venkateswara College of Engineering and Technology, India
[2,3,4]Department of Electrical and Electronics Engineering, Adhi College of Engineering and Technology, India

Vijayabrabhu85@gmail.com, jaibhaskar15@gmail.com

Abstract. Increasing in the power demand for day to day life, apart from fossil fuels wind energy conversion system (WECS) contributes largely, small wind turbines (SWT) and off shore installations become popular and with the continual research newly designed efficient generators and power converter design new wind turbine models are expected to launch in the recent years. At present WECS operates at fixed speed and variable speed drives using induction generator principle, such that Squirrel cage induction generators (SCIG) and Doubly fed induction generators have a geared operation, Permanent magnet synchronous generator (PMSG) goes gearless. In elaborate the comparisons, advantages and drawbacks of various WECS is discussed in this work.

Keywords: WECS, DFIG, SCIG, PMSG.

1. Introduction

Exponentially human population rise at an enormous number leads to more demand in energy [1]. To fulfil the requirement power generation becomes necessary and essential thing for national growth also. Energy consumption and development have a same state ofart. Hence progressive power production is much awaited[2]. Dependency of fossil fuels for energy production leads to emission of pollutant gases like carbon mixturehave a poisonous effect on natural world. Thereby, possible way ofproducing energy utilizing renewable sources like solar, wind will not emit hazardous gases[3]. In recent days wind is gaining more attentiveness due to its infrastructure, technological advancements and cost efficient[4].

In the developing countries almost one-third of human life's proceeding their lives without power supply, since wind turbines is there for almost bicentennial to generate power, WES is a cost-effective mechanism for power demand[5][6]. Environmental friendly and being lucrative WES is considered as a precious source for energy production[7]. Even though wind attainability varies from place to place, it lacks in continual power supply[8]. Due to essential heavy requirement of power the turbine technology have a considerable change from the recent past[9]. Continual improvement in wind turbine modelling technology in the areas concern power converters, streamlined flowing sleek and mechanical power plant design approaches it an powerful source of energy[10]. In remote areas usually wind turbines are installed along with wind farms where power supply from grid is insufficient and the wind farm since energy supply from grid is insufficient and the produced energy from the farm is normally routed to storage[11][12]. For additional power requirement there exists a direct connection to grid from wind plants [13].

In the earlierdecades, horizontal axis wind turbine (HWAT) have largely been used to produce electricity with higher capacity for power demand[14]. Since the major drawback in HWAT system is the power loss due to the use of gear mechanism which leads to huge maintenance cost, all these factors made the overall HAWT system costlier[15]. To overcome this issue Vertical axis wind turbine (VAWT) were introduced in early 19th century, However, VAWT resolved the issue by replacing the gear box to direct drive generator system which minimizes the maintenance cost by reducing power losses[16].

This paper is classified in to different sections. Section 2 discusses the different arrangement and numerous designs of a wind turbine. Next chapter 3 elaborates commonly used generators listed as Induction, Synchronous, Double Fed Induction (DFIG), Squirrel Cage Induction (SCIG) and Permanent Magnet Synchronous (PMSG). In Section 4 will enlighten the comparison of different generator topologies. Section 5 engages to the selection of best suitable generator as per the power demand.

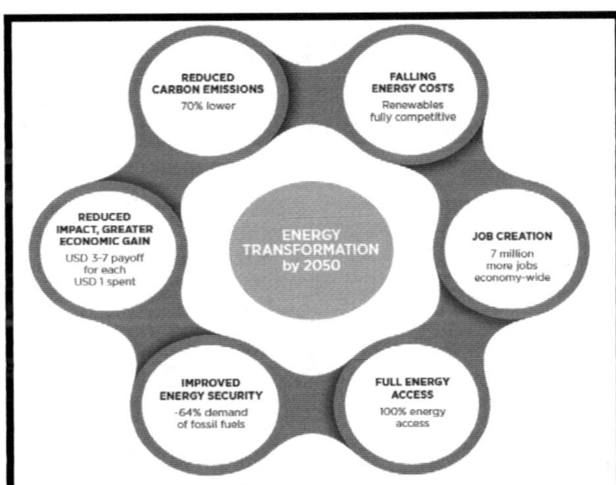

Figure 1. Predicted transformation in near future

2. WIND TURBINE DESIGN AND TOPOLOGIES

Fixed and variable speed drive systems are commonly used wind turbine topologies in general[17]. In recent decades, wind turbine technology have rapid growth from older fixed speed WT to variable speed WT, mainly 1kw SCIG type are available[18]. Variable speed WT intended to aim maximum aerodynamic efficiency over a different range of wind speeds. Compared with horizontal axis wind generator with 3 rotor blades and mechanical parts and heavy generator mounted at the nacelle. The nacelle is placed very high up on a tower. The blades of large scale WTS diameters to a rotational speed between 5 and 25 rpm. Nowadays variable speed WT with pitch control systems replacing later one. Fig 2. shows that wind turbine systems have the potential to transform kinetic (mechanical) energy of the wind into power[19]. The conversion process is derived from the following equations.

$$P_M = C_P(\lambda, \beta)P_W \qquad (1)$$

$$P_W = \frac{1}{2}\Pi R^2 V^3 \rho_{AIR} \qquad (2)$$

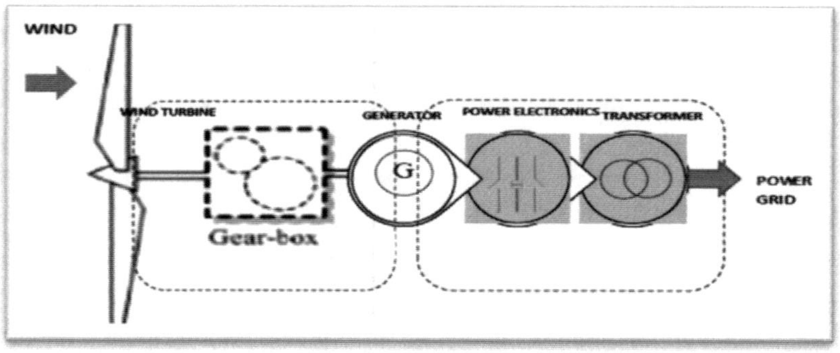

Figure 2. Power Generation from Wind energy system

Where P_W and P_M denotes the wind and mechanical power of the wind. Performance coefficient is referred as C_P of the turbine, R represents the radius of the rotor,and the wind speed V and air density represents as ρ_{AIR}.In $C_P(\lambda,\beta)$ coefficient is in function of λ that is the tip speed ratio (TSR) and β refers the pitch angle of the blade in degrees. Cp is at its peak value when $\beta=0$ and $\lambda=8.1$ theT_{mec} (mechanical torque) can bedetermined in the below relationship, where torque coefficient (C_m) can be written as.

$$C_M = C_P/\lambda \qquad (3)$$

$$T_{mec} = \frac{1}{2}\Pi R^2 V^3 \rho_{AIR} \, (4)$$

3. Various Types of Generators

3.1.Induction generator (IG)

The principle of induction motor to produce electric power is adopted by the induction generator (IG) and synchronous generator is of alternating type. IG rotates the rotor above the synchronous speed by mechanically turning their rotor. In generator working, turbine or engine rotates the rotor far beyond the synchronous speed (negative slip). The stator flux induces currents in the rotor, but since the opposing rotor flux is now cutting the stator coils, an active current is produced in stator coils and the motor now operates as a generator, sending power back to the electrical grid.

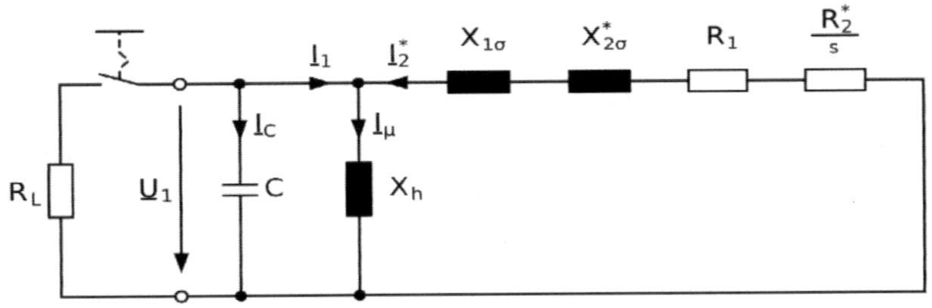

Figure 3. Induction generator (IG) equivalent modelling

3.1.1. Application and advantages Wind turbines, small wind turbine and micro hydro installations are often uses Induction generators (IG) which have the potential to generate useful power at varying rotor speeds. Comparison to the other generator types IG are mechanically and electrically simpler. Even which it is also much rugged, hence it need not require brushes or commutators.

3.2. Synchronous generator (SG)

One of the generator used for variable speed turbine application is synchronous generator, it can be a suitable selection for variable speed operation since it has low rotational synchronous speed that produces the voltage at grid frequency.it does not require a pitch control mechanism since cost cutting of turbine is possible and occur stress on generator and turbine. SG will generate variable voltage and variable frequency power. Utilizing automatic voltage regulator (AVR) to excite field voltage so as the output voltage of the SG can be controlled.

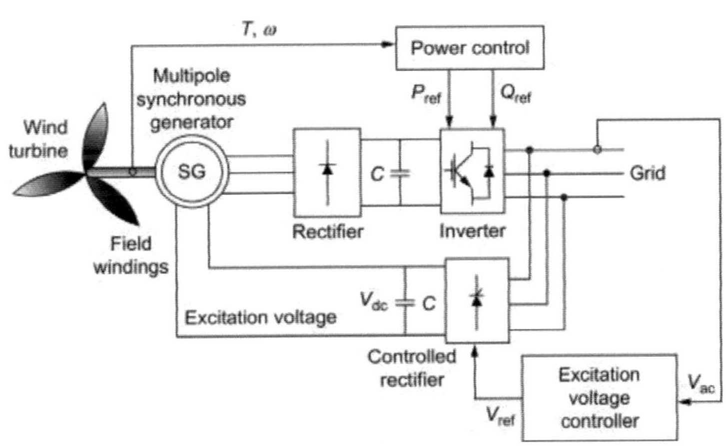

Figure 4. Schematic of synchronous generator

3.3. DFIG (Doubly Fed Induction Generator)

Another popular series in this type of generator is DFIG, maximum energy is captured in variable speed winds through which power electronic interface controls the rotor currents, the matching between the rotor and generator speed ranges should be achieved by a coupled gearbox. Since the process of the rotor is driven by the power electronics switches it captures less than quarter of the complete output power. The power switching converter is to make conversion from AC to DC to AC usually of variable frequency drive.

Normally the soft switches with high stability like IGBT converters are made: A DC-link connects the grid side converter and rotor side converter. The decoupling between the grid frequency and rotor mechanical frequency is also achieved by the power switching converter which progresses variable speed operation. Since the rotor power is from the power switching converter, The Rotor side power converter topology holds the control of active and reactive powers with controlled harmonics at the generator,at the Grid side power converter topology maintains the power factor to high enough. The main advantage of DFIG is reduced power losses with minimal cost is achieved. The working representation of DFIG is as below in Fig.5.

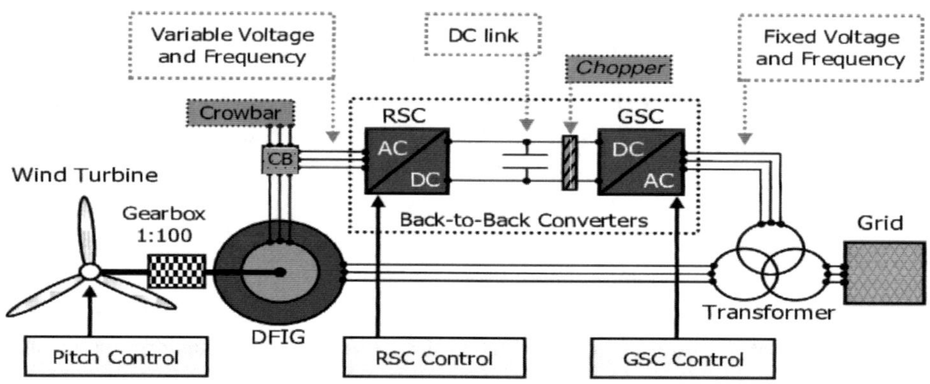

Figure 5. RSC and GSC Representation in DFIG

3.3.1. DFIG Advantages

- Compared with the earlier generator type it is simpler in mechanical design and electrical structure.
- Seam is Rugged and brushless produces higher efficiency and higher energy.
- Three stage geared DFIG is weightless for cost effective solution,
- Power switching converter topologies matches the reactive power and ensures smooth integration of grid.
- Power Conversion Rating is only about 25%-30% in DFIG as compared to 100 % of maximum true power of the generator, it possible Can reach to about 30% of synchronous speed, thus have a wide speed range.

3.3.2. DFIG Disadvantages

- In these type, gearboxes are still essential, prior in multiple pole DFIG achieving lower speed is technically not possible.
- Grid side faulty conditions are not simpler to assemble
- Since produces upper than lower reliability and have limited life because of gear faults and bearing fault.

3.4. SCIG (Squirrel cage Induction Generator)

The SCIG is an strongly a better consideration in the category of variable drive wind power turbine and SCIG is in fully coordination with the power converter to adjust the voltage generated and frequency for load side in an standalone loads. Extraction of more power from wind is limited since it causes overload at the generator. So as to grasp an optimal power, blade pitch angle is required, Since SCIG prolongs as a replacement energy resource by having simpler design structure, requires much lesser maintenance, excellent power to weight ratio so it is robust.

3.4.1. SCIG Advantages

- The construction is sturdy and simpler mechanism makes this machine very popular.
- Brushes are not required in SCIG compared to DFIG.
- Rotor bars are designed to be impervious to quivering smut
- Suitable for variable speed operations since grid is absolutely decoupled.
- Protection from the grid side is higher to avoid short circuit power to the generation side.

3.4.2. SCIG Disadvantages:

- Grid side and load side converters are required for operating.
- Compared with DFIG for variable speed operation it requires more number of power converter switches which makes more switching losses.
- SCIG is not suitable for gearless operation in multiport.

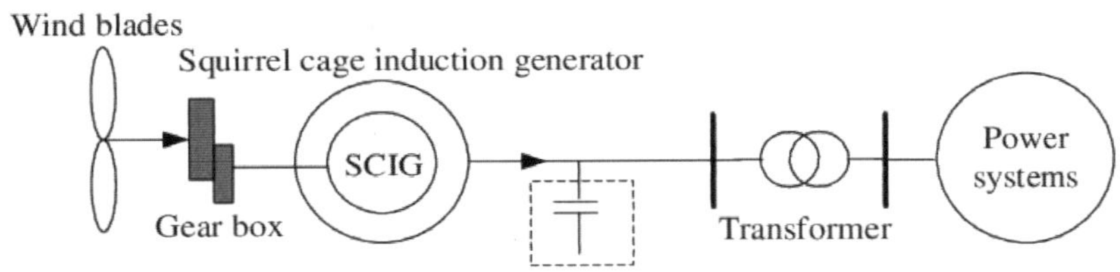

Figure 6. SCIG working model blocks

3.5. PMSG (Permanent magnet synchronous generator)

PMSG drive does not require gearbox since it does not have a rotor current with minimum rotational speed, since the drive is gearless small wind turbines are easily possible at consumer side with high efficiency at low maintenance cost. The principle of pmsg is that it generates torque from wind power. The developed torque is shifted to the rotor shaft of the generator. The mechanical rotation of the rotor acceleration depends on the produced electrical torque from the generator difference to the mechanical torque.

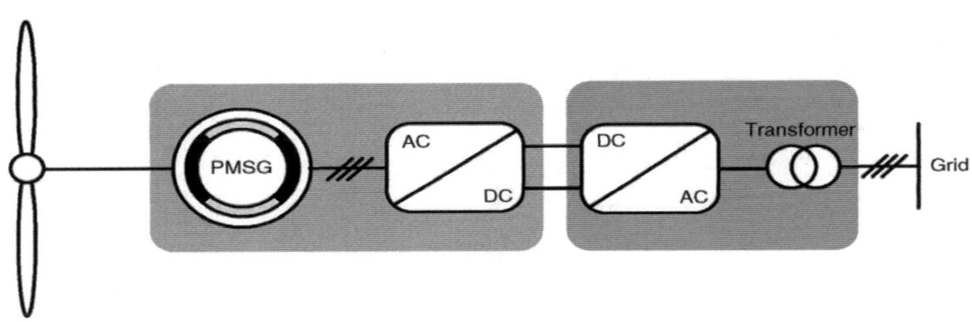

Figure 7. PMSG structure representation

IOP Conf. Series: Materials Science and Engineering 1177 (2021) 012004 doi:10.1088/1757-899X/1177/1/012004

3.5.1.PMSG Advantages:

- Maintenance cost will cut down due to no gearbox.
- Possibility of faults in the generators will be eliminated since there is no gears and bearings, thus produces higher reliability.
- Minimum weight that produces higher energy yield.

3.5.2. PMSG Disadvantages:

- Compared with the SCIG, the outer diameter of the PMSG is almost double in size.
- It is a new technology drive, the cost being higher.
- Wind turbine capacity over 4MW leads to a critical proportions and have increased mass and weight.

Table 1 Comparison of Different Generator Topologies

Generator Type	STRENGTHS	WEAKNESS
Induction Generator	• Maximum speed capacity range • Generator do not require brushes • Reactive and active power are in full control	• power converter is must • Gears are unavoidable
Synchronous Generator	• Maximum speed range • Gear can be avoided • Reactive and active power are of complete control	• Compact converter for field is enough • AC/DC/AC power converter is required • Multipole generator in case of direct driven topology
Doubly-fed Induction Generator	• Speed range limited from -30% to 30% around synchronous speed • PWM inverter are of small capacity and inexpensive • Reactive and active power are of complete control	• Slip rings are unavoidable • Gears required
Permanent Magnet Synchronous Generator	• Maximum speed range • Gear can be avoided • Reactive and active power are of complete control • No Brushes ensures low maintenance • Field has no power converter	• AC/DC/AC power converter is required • Multipole generator (big and heavy) • Permanent magnets are necessary

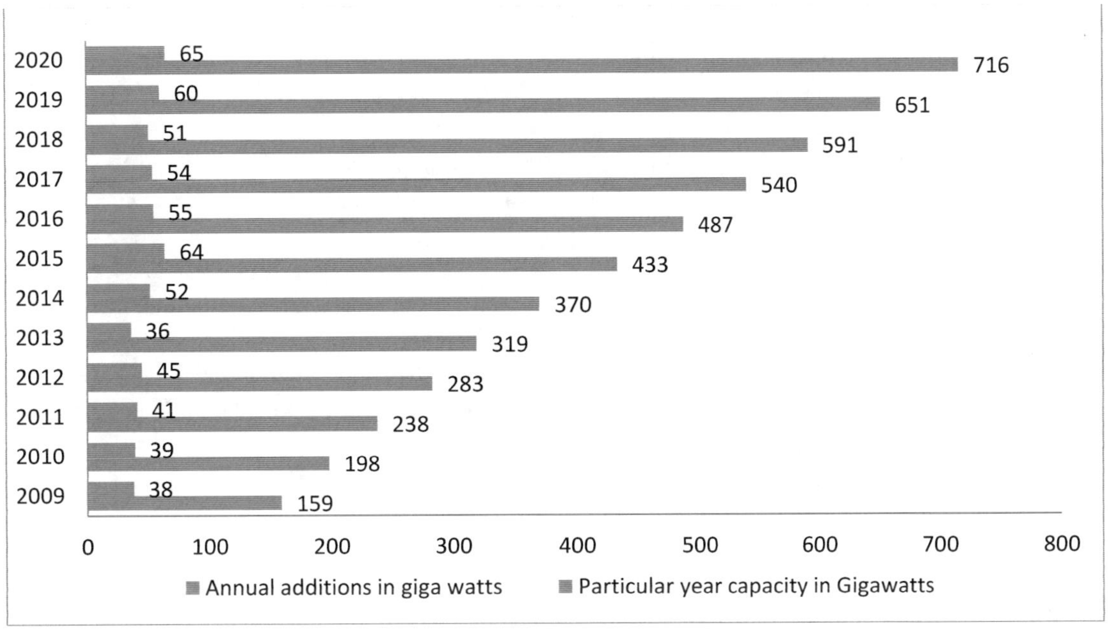

Figure 8. Wind energy power global and annual additions, around 60 GW increased till 2019.

4. Conclusion

This paper has intended to review the various technology forgenerators used in WECS. Due to the fact that the variable speed machines are in demand that produces lower mechanical stress and higher power capture. DFIG plays a key role in the current market by comparing with the variable speed wind turbine DFIG have the advantage of only 30% of the power generated passes through the converter topology this factor leads to a sensible cost cutting advantage compared with the cheap cost of nearby converters and power switches. Grid faults occurrence possibility is more in DFIG due to large peak currents, and SCIG suffers to a short circuit problems at grid. The parameters consideration of overall efficiency, long lasting and reliability and market availability, the PMSG wind turbine generator without the gearbox leads the top. Since they are larger in size, can be still installed at offshore where there is enough wind speed.

5. References

[1] N. Abas, A. Kalair, and N. Khan, "Review of fossil fuels and future energy technologies," *Futures*, vol. 69, pp. 31–49, 2015, doi: 10.1016/j.futures.2015.03.003.

[2] G. P. Hammond, "Energy, environment and sustainable development: A UK perspective," *Process Saf. Environ. Prot.*, vol. 78, no. 4, pp. 304–323, 2000, doi: 10.1205/095758200530826.

[3] Alagumariappan, Paramasivam, Najumnissa Jamal Dewan, Gughan Narasimhan Muthukrishnan, Bhaskar K. Bojji Raju, Ramzan Ali Arshad Bilal, and Vijayalakshmi Sankaran. 2020. "Intelligent Plant Disease Identification System Using Machine Learning" *Engineering Proceedings* 2, no. 1: 49. https://doi.org/10.3390/ecsa-7-08160

[4] S. Breukers and M. Wolsink, "Wind power implementation in changing institutional landscapes: An international comparison," *Energy Policy*, vol. 35, no. 5, pp. 2737–2750, 2007, doi: 10.1016/j.enpol.2006.12.004.

[5] G. Antonucci *et al.*, "Designing a Model Landscape for Multiple Values : Envisioning a

Sustainable Future for Middlebury College Lands," no. 401, 2017.

[6] S. Sezgin, "The Third Industrial Revolution: How Lateral Power is Transforming Energy, the Economy, and the World," *Turkish J. Bus. Ethics*, vol. 11, no. 1, 2018, doi: 10.12711/tjbe.2018.11.1.0005r.

[7] P. Maegaard, "Wind energy development and application prospects of non-grid-connected wind power," *WNWEC 2009 - 2009 World Non-Grid-Connected Wind Power Energy Conf.*, pp. 1–3, 2009, doi: 10.1109/WNWEC.2009.5335847.

[8] K.B. Bhaskar, T.S. Sivakumaran and M. Devi ' comparison of a novel 15 level cascaded asymmetric multicell inverter using pi and fuzzy logic controller', International Journal of Pharmacy & Technology issue 3 volume 8 sep -16 (scopus)

[9] J. Kabouris and F. D. Kanellos, "Impacts of large-scale wind penetration on designing and operation of electric power systems," *IEEE Trans. Sustain. Energy*, vol. 1, no. 2, pp. 107–114, 2010, doi: 10.1109/TSTE.2010.2050348.

[10] E. Echavarria, B. Hahn, G. J. W. Van Bussel, and T. Tomiyama, "Reliability of wind turbine technology through time," *J. Sol. Energy Eng. Trans. ASME*, vol. 130, no. 3, pp. 0310051–0310058, 2008, doi: 10.1115/1.2936235.

[11] S. B. A. Kashem, M. E. H. Chowdhury, A. Khandakar, J. Ahmed, A. Ashraf, and N. Shabrin, "Wind power integration with smart grid and storage system: Prospects and limitations," *Int. J. Adv. Comput. Sci. Appl.*, vol. 11, no. 5, pp.552–569, 2020, doi: 10.14569/IJACSA.2020. 0110570

[12] Bhaskar, K. B., & Alagumariappan, P. (2020). Energy Channeling Led Driver Technology to achieve Flicker Free Operation with Zeta Converter for Power Quality Improvement. *International Journal of Automation and Smart Technology*, *10*(1), 401-407.

[13] Z. Chen, "Issues of connecting wind farms into power systems," *Proc. IEEE Power Eng. Soc. Transm. Distrib. Conf.*, vol. 2005, pp. 1–6, 2005, doi: 10.1109/TDC.2005.1547160.

[14] R. Kumar, K. Raahemifar, and A. S. Fung, "A critical review of vertical axis wind turbines for urban applications," *Renew. Sustain. Energy Rev.*, vol. 89, no. February, pp. 281–291, 2018, doi: 10.1016/j.rser.2018.03.033.

[15] K.B. Bhaskar and T.S.Sivakumaran 'Implimentation of 11 level cascaded multilevel inverter using level shifting pulse with modulation technique with different loads', International journal of electrical engineering(IPASJ),volume 2,issue 10 Nov 2014

[16] Paramasivam A., Bhaskar K.B., Madhanakkumar N., Vanchinathan C. (2020) Analysis of an Enhanced Positive Output Super-Lift Luo Converter for Renewable Energy Applications. In: Siano P., Jamuna K. (eds) Advances in Smart Grid Technology. Lecture Notes in Electrical Engineering, vol 687. Springer, Singapore. https://doi.org/10.1007/978-981-15-7245-6_11

[17] D. Song, J. Yang, Z. Cai, M. Dong, M. Su, and Y. Wang, "Wind estimation with a non-standard extended Kalman filter and its application on maximum power extraction for variable speed wind turbines," *Appl. Energy*, vol. 190, pp. 670–685, 2017, doi: 10.1016/j.apenergy.2016.12.132.

[18] S. H. Qazi and M. W. Mustafa, "Review on active filters and its performance with grid connected fixed and variable speed wind turbine generator," *Renew. Sustain. Energy Rev.*, vol. 57, pp. 420–438, 2016, doi: 10.1016/j.rser.2015.12.049.

[19] P. K. Keung, P. Li, H. Banakar, and B. T. Ooi, "Kinetic energy of wind-turbine generators for system frequency support," *IEEE Trans. Power Syst.*, vol. 24, no. 1, pp. 279–287, 2009, doi: 10.1109/TPWRS.2008.2004827.

Design and Implementation of Hybrid Asymmetric 15-Level Cascaded Modified H-Bridge Multilevel Inverter(CMHB-MLI) to reduce the Total Harmonics Distortion

K B Bhaskar[1] A Vijayaprabhu[2] R Kumaresan[3] E Kokila[4]
[1,3,4]Department of Electrical and Electronics Engineering, Adhi College of Engineering and Technology, India
[2]Assistant Professor, Department of Electronics and Communication, Sri Venkateswara College of Engineering and Technology, India

jaibhaskar15@gmail.com[1], Vijayabrabhu85@gmail.com[2]

Abstract. This new cascaded modified H-bridge asymmetrical multilevel inverter (CMHB-MLI) module proposes a 15 levels with three unequal dc sources and 10 switches. The design approaches with preferable advantage with greater quality comparatively to similar modules with minimum number of semiconductor switches utilizing lower switching frequency and less dc sources.Opposite phase disposition pulse width modulation (OPD-PWM) switching technique is utilized for controlling the Semiconductor switches in CMHB-MLI. In asymmetrical topology the input source are designed with unequal magnitude is done deliberately with binary nature of voltages such as 10Vdc, 20Vdc and 40Vdc. In asymmetrical designing of MLI produces a number of different output levels with the similar number of power semiconductor switches.The results are designed using MATLAB and hardware implementation validates improved quality performance with lower total harmonicsdistortion without utilizing any filtering circuit in the output with reduced switching stress.

Key words – Asymmetric, Hybrid, CMHB-MLI, opposite phase disposition pulse width modulation (OPD-PWM), Total harmonic distortion.

1. Introduction

In the past decades, multilevel inverters had a great attention towards cost effective systems with wider range of applications mainly in the areas of power quality improvement devices, flexible AC transmission systems, larger motor drives and green energy converters[1]–[3].The anxiety to ozone layer depletion caused by burning of fossil fuels for electricity generation leads to usage of natural available green resources such as photovoltaic, wind, geothermal etc., to improve the efficiency of the system and cost cutting a power electronic conversion switching technology is a valid requirement towards power generation from solar and wind[4]–[6].

Compared to the conventional inverter, MLIs have numerous advantages that produces common mode voltage with reducing switching stress by drawing lowering input current with low distortion, MLIs can operate at lower switching frequencies and achieves higher efficiency with lower switching stress[7], [8], By proper PWM technique selective harmonics can be eliminated which can be resulted in lower total harmonic distortion in the output waveforms without usage of any filter[9]–[12].

In general, the MLI are classified as Neutral point diode clamped (NPDC), Flying capacitor (FC) and Cascaded H-Bridge (CHB). In NPDC multilevel inverter makes use of diodes which transfers minimum amount of voltage leads to drawback of producing a maximum output voltage is half of the input voltage[13]–[16]. In FC multilevel inverter approaches to more number of capacitors are connected in series which causes switching redundancy impact to reduction in voltage level and voltage balancing is difficult thereby produces half output compared to the input voltage[17]–[19]. CHB multilevel inverter avoiding usage of bulky transformer, clamping diodes and capacitor which required in NPDC and FC and each cell in an H-bridge produces three different voltages like zero, positive DC and Negative DC voltages, the main advantage of multilevel inverteris that it requires less components[20]–[23].

Making concern of the vital factor of MLI design approach for high voltage applications will generate increased stress on output H-bridge therefore it is necessary to build a new topology which will make stress even to all the switches[24]–[27].

The proposed asymmetric CMHB-MLI aims to achieve lowering switching stress to the switches by designing a proper pulse dispositioning scheme and achieves lower total harmonics distortion with greater efficiency at the output voltage level and cost efficient, and also targeted to produce 15 level with less number of switches.

The paper proceeds with, section II describes the symmetrical and asymmetrical cascaded H Bridge multilevel inverter and analysis of conduction and switching losses, section III elaborates the proposed CMHB-MLI circuit description with switching state analysis with the analysis of operating modes of converter, section IV validates the simulation and hardware results and section V concludes the proposed CMHB-MLI continued with the references.

2. Symmetrical and Asymmetrical Cascaded H Bridge Multilevel Inverter

In symmetrical MLI with 'n' equal dc voltage offer 2n+1 distinct voltage level at the phase output. Since the required number of devices depends on output voltage level, the problem such as increasing inverter size, cost, and installation area and control complexity should be avoided.

To overcome the above mentioned problem an asymmetrical MLI is approached for providing a large number of output voltage levels without increasing the number of bridges, a method of choosing non identical DC input sources is referred as binary configuration. The topology can achieve 2^{n+1}-1 distinct voltage levels with 'n' cascaded H-bridges, this approach gets better performance in terms of levels than symmetrical MLI topology with the same number of switching components and DC buses. However, high variability in the consequent DC voltage magnitudes leads to charge balance problem as well as the higher voltage stress. The general representation for Asymmetrical model as,

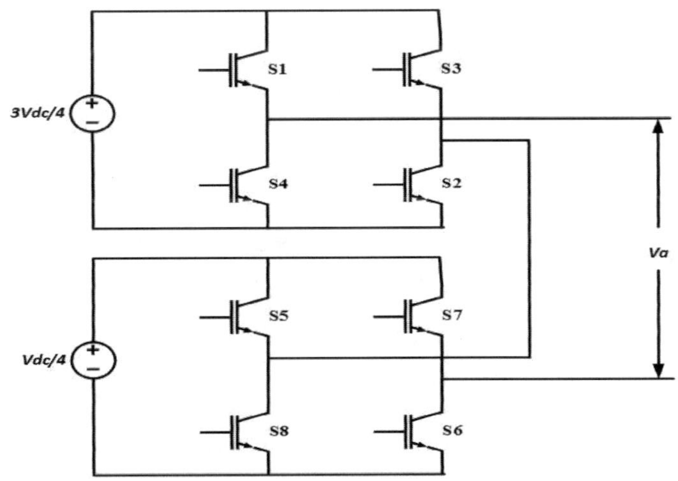

Figure 1. Basic structure of H-bridge multilevel inverter

2.1 Conduction loss

Usually losses will be calculated in two types conduction and switching loss. First conduction loss, developed to multilevel inverter is the instantaneous conduction losses of a transistor, the losses that occur during the power device is the on-state and conducting current, therefore, power dissipation during conduction is computed by multiplying on state saturation voltage and on state saturation current and it is given as.

$$p(t) = V(t) * I(t) \quad (1)$$

In the equation 1, V(t) represents on state saturation voltage and I(t) is the load current, and the instantaneous conduction losses of semiconductor switches $P_{CS}(t)$ is given in equation 2.

$$P_{cs}(t) = [V_s + R_s i^{\beta(t)}]i(t) \quad (2)$$

The average conduction loss of main circuit is represented in equation 3 as,

$$P_{c,Mc} = \frac{1}{\pi}\int_0^{\pi}[N_s(t) + P_{cs}(t)]d\omega(t) \quad (3)$$

The output current of auxiliary circuit to be sinusoidal as i (t) = I_MSinωt and conduction losses of auxiliary circuit can be calculated as below.

$$P_c A_c = \frac{1}{\pi}\int_0^{\pi}[x(t)V_{T,j} + x(t)R_{T,j}i(t)]i(t).d\omega t$$

$$= \frac{1}{\pi}\int_0^{\pi} 2P_{cs}(t).d\omega t$$

$$= \frac{2}{\pi}[V_s + R_s i^{\beta}(t)]i(t).d\omega t$$

$$P_c A_c = \frac{2}{\pi}[2V_{TH} I_M + R_{TH} I_M{}^{\beta H} \int_0^{\pi} \sin^{\beta+1} \omega t . d\omega t \quad (4)$$

2.2 Switching losses

Switching loss is the power dissipation during turn-on and turn-off switching transitions, the switching loss can be determined by plotting the voltage and current waveform in controllable switch during the switching transition and multiplying the waveforms point by point to get an instantaneous power waveform, and the energy loss during turn-off of a switch by using linear approximation can be expressed as,

$$E_{off}, k = \int_0^{toff} V(t) I(t).dt \quad (5)$$

$$= \int_0^{toff} [(\frac{V_{offk}}{t_{off}} t)(-\frac{I}{t_{off}}(t - t_{off}))]dt \quad (6)$$

$$E_{off}, k = \frac{V_{off}, k, It_{off}}{6} \quad (7)$$

Similarly

$$E_{on}, k = \int_0^{ton} V(t) I(t).dt \quad (8)$$

$$= \int_0^{ton} [(\frac{V_{onk}}{t_{on}} t)(-\frac{I}{t_{on}}(t - t_{on}))]dt$$

$$E_{on}, k = \frac{V_{on}, k, It_{on}}{6} \quad (9)$$

Average switching losses can be calculated by using equation 10.

$$P_{sw} = 2f[\sum_{k=1}^{N_{switch}} (\sum_{i=1}^{N_{on,k}} E_{on}, K_i + \sum_{i=1}^{N_{off,k}} E_{off} K_i)] \quad (10)$$

And the Total power losses will be estimated as the combination of conduction losses, auxiliary switch conduction loss and average switching loss will be expressed in the equation 11.

$$P_{loss} = P_{c,Mc} + P_{c,Ac} + P_{sw} \quad (11)$$

3. Structure of the proposed CMHB-MLI

The proposed CMHB-MLI is shown in the Fig. 2, is a cascaded unit of 15 level inverter, for its dc-dc converter section that consists of number n semiconductor section, it is capable of producing n+1 voltage levels according to different switching states. With the operation of the H-bridge, a total of 2n+3 voltage level can be produced.

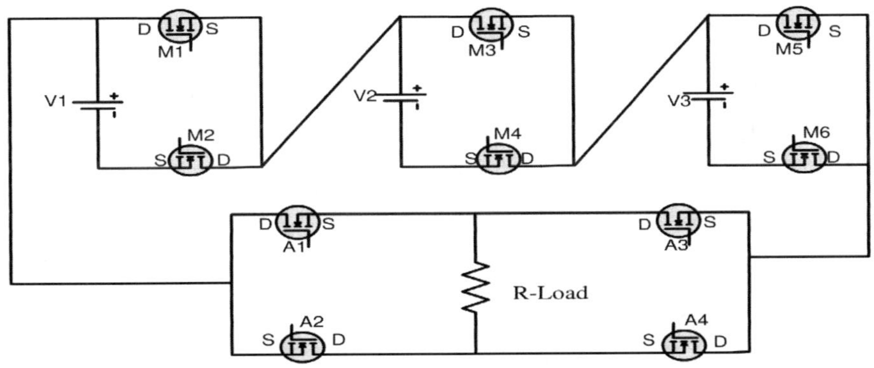

Figure 2. Proposed CMHB-Multilevel inverter topology

3.1 Circuit Description and state analysis

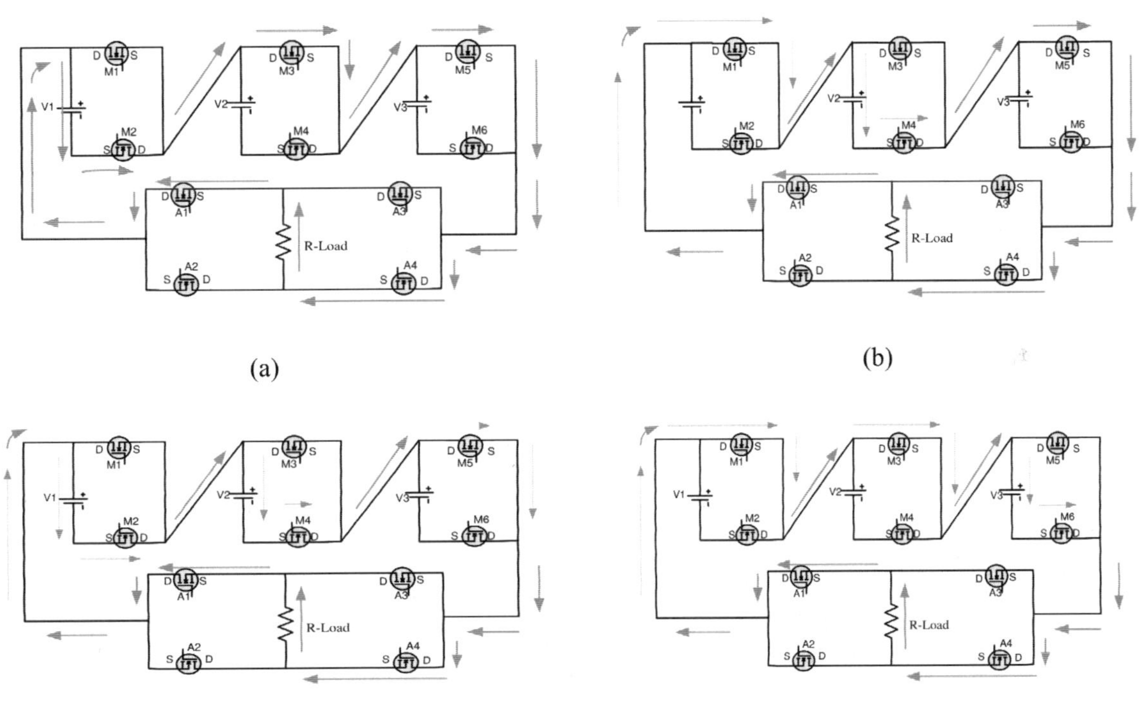

(a) (b)

(c) (d)

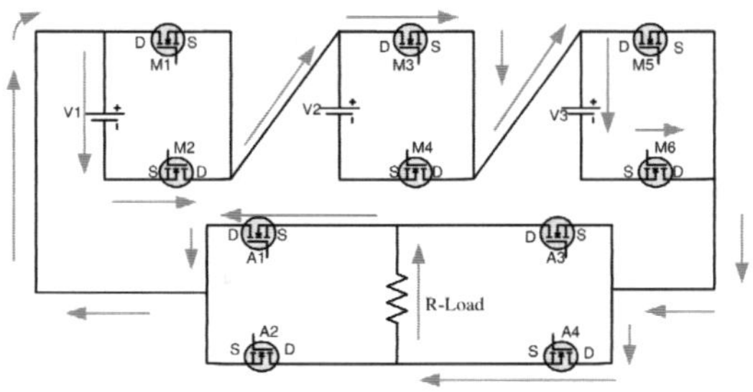

(e)

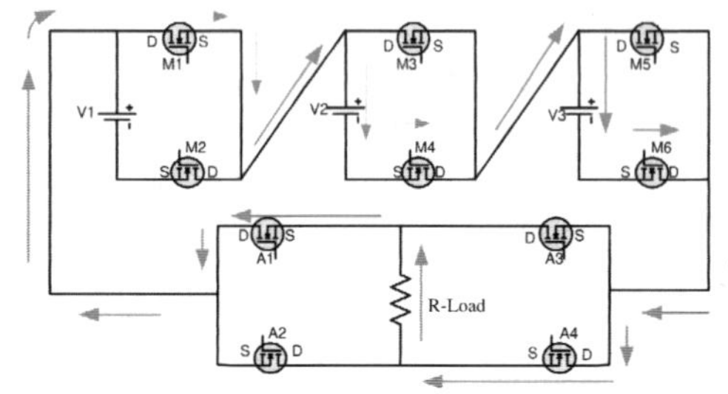

(f)

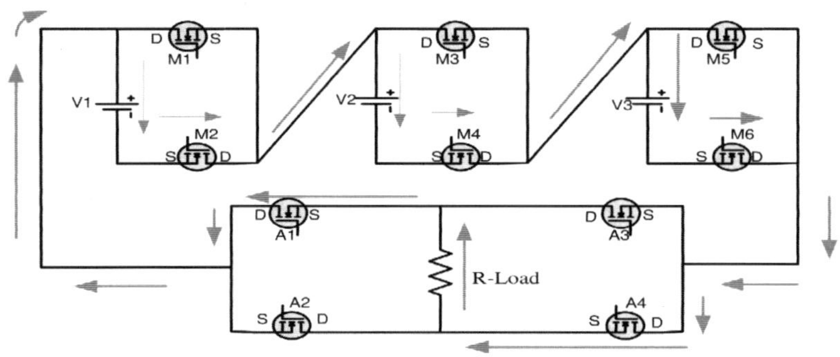

(g)

Figure 3. Working states for the Proposed CMHB-Multilevel inverter topology. (a) maximum positive output voltage (+Vdc) (b) +6Vdc/7 (c) +5Vdc/7 (d) +4Vdc/7 (e) +3Vdc/7 (f) +2Vdc/7 (g) +Vdc/7

Table 1. Switching States for the Proposed Inverter

output voltage	**M1**	**M2**	**M3**	**M4**	**M5**	**M6**	**A1&A4**	**A2&A3**
70	1	0	1	0	1	0	1	0
60	1	0	1	0	0	1	1	0
50	1	0	0	1	1	0	1	0
40	1	0	0	1	0	1	1	0
30	0	1	1	0	1	0	1	0
20	0	1	1	0	0	1	1	0
10	0	1	0	1	1	0	1	0
0	0	0	0	0	0	0	0	0
-10	1	0	1	0	1	0	0	1
-20	1	0	0	1	0	1	0	1
-30	1	0	0	1	1	0	0	1
-40	0	1	1	0	0	1	0	1
-50	0	1	1	0	1	0	0	1
-60	0	1	0	1	0	1	0	1
-70	0	1	0	1	1	0	0	1

3.2 Operating modes of the proposed 15 level CMHB Multilevel inverter

The positive staircase output produced by the proposed switching pattern is explained from maximum positive voltage (Vdc) to zero level, the voltage level obtained is Vdc, 6Vdc/7, 5Vdc/7, 4Vdc/7, 3Vdc/7, 2Vdc/7 and Vdc produces the positive output is explained below.

(i) When MOSFET switch M1, M3, M5 is turned ON, and switches in H-bridge A1 and A4 is at on position, and the switch M2, M4, M6, A2 AND A3 is at OFF position the current flows through from the source and produces a maximum positive output voltage of (Vdc) of the load.

(ii) When the MOSFET switch M1, M3, M6 is turned ON, and switches in H-bridge A1 and A4 is at on position, and the switches M2, M3, M5 is at OFF position the current flows through from the positive potential and produces a positive output voltage of (6Vdc/7) at the load.
(iii) When the MOSFET switch M1, M4, M5 is turned ON, and switches in H-bridge A1 and A4 is at on position, and the switches M2, M3, M6 is at OFF position the current flows through from the positive potential and produces a positive output voltage of (5Vdc/7) at the load.

(iv) When the MOSFET switch M1, M4, M6 is turned ON, and switches in H-bridge A1 and A4 is at on position, and the switches M2, M3, M5 is at OFF position the current flows through from the positive potential and produces a positive output voltage of (4Vdc/7) at the load.

(v) When the MOSFET switch M2, M3, M5 is turned ON, and switches in H-bridge A1 and A4 is at on position, and the switches M1, M4, M6 is at OFF position the current flows through from the positive potential and produces a positive output voltage of (3Vdc/7) at the load.

(vi) When the MOSFET switch M2, M3, M6 is turned ON, and switches in H-bridge A1 and A4 is at on position, and the switches M1, M4, M5 is at OFF position the current flows through from the positive potential and produces a positive output voltage of (2Vdc/7) at the load.

(vii) When the MOSFET switch M2, M4, M5 is turned ON, and switches in H-bridge A1 and A4 is at on position, and the switches M1, M3, M6 is at OFF position the current flows through from the positive potential and produces a positive output voltage of (Vdc/7) at the load.

Similar to the positive staircase output obtained by different switching pattern, negative staircase output voltage from -Vdc/7, -2Vdc/7, -3Vdc/7, -4Vdc/7, -5Vdc/7, -6Vdc/7 and –Vdc also can be obtained by reversing the pattern of switching at each level and 15-level output is obtained.

IOP Conf. Series: Materials Science and Engineering 1177 (2021) 012005 doi:10.1088/1757-899X/1177/1/012005

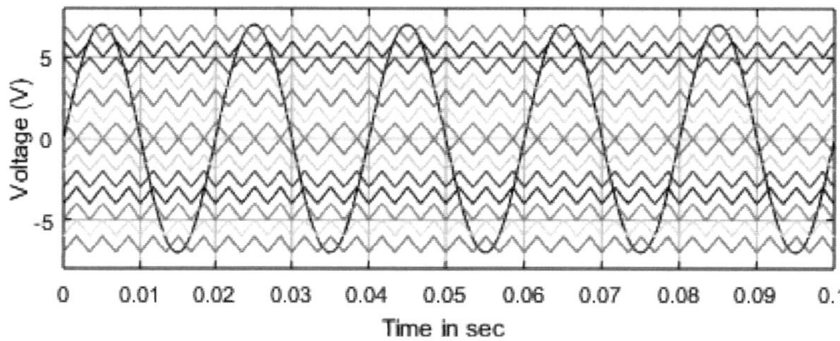

Figure 4. Phase disposition pulse width modulation (OPD-PWM) switching technique

4. Simulation and Hardware Results

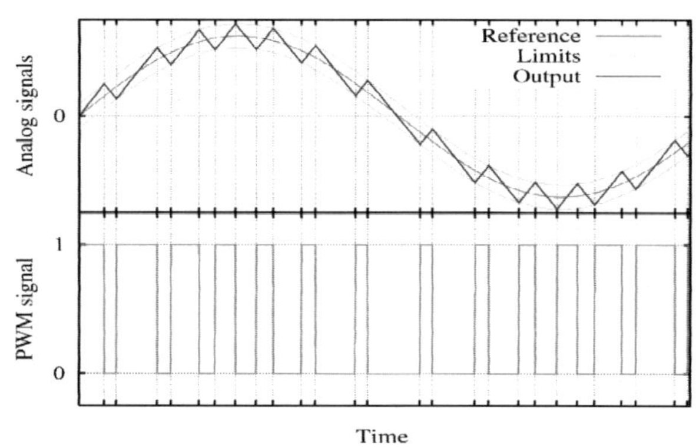

Figure 5. Pulse width modulation

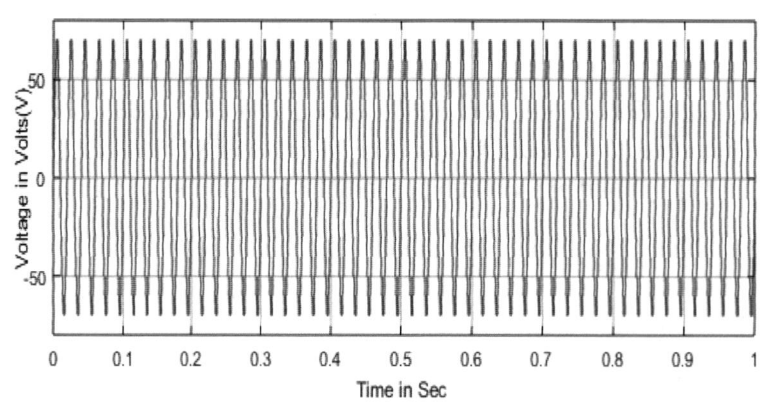

Figure 6. PWM voltage

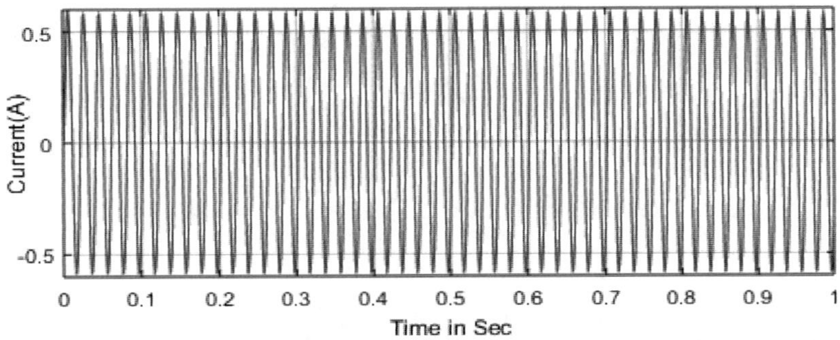

Figure 7. PWM current

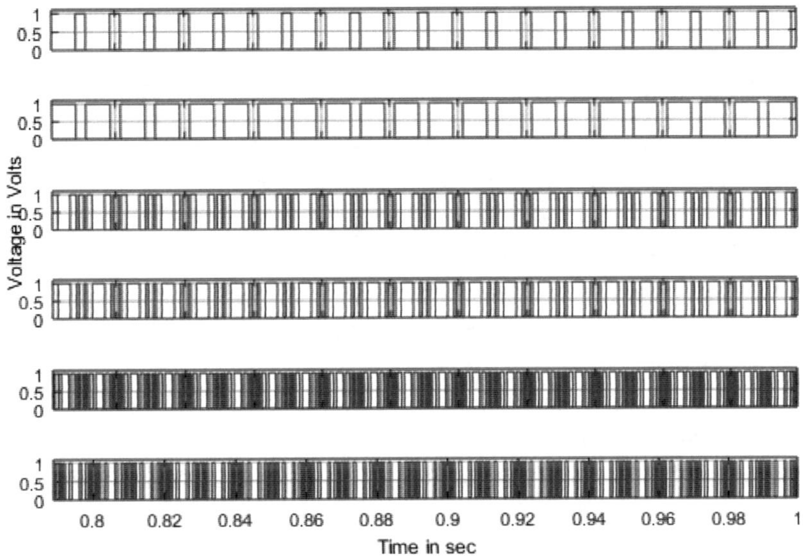

Figure 8. PWM technique

IOP Conf. Series: Materials Science and Engineering 1177 (2021) 012005 doi:10.1088/1757-899X/1177/1/012005

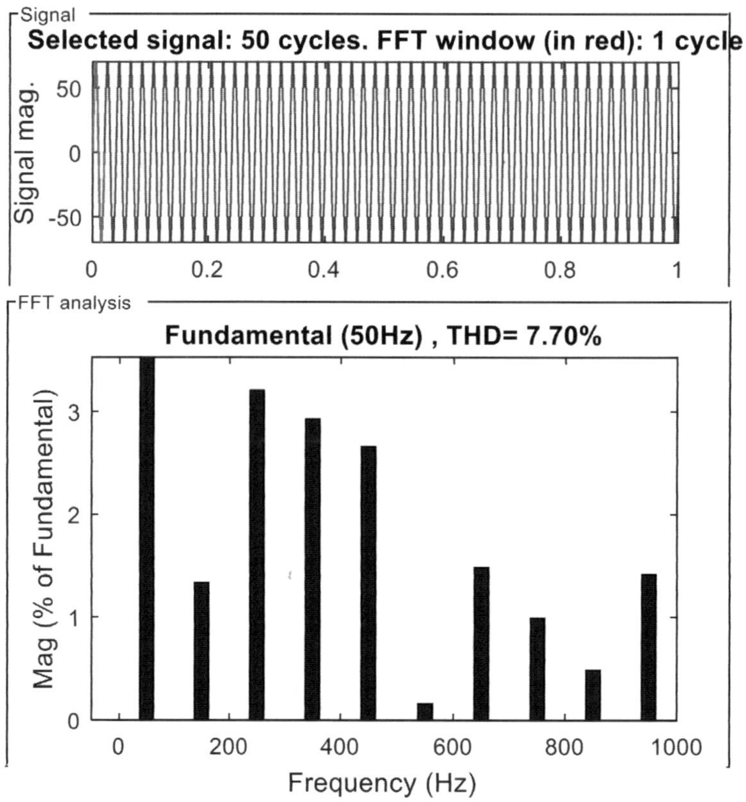

Figure 9. Total harmonic distortion

In above fig we calculated THD value in 15 level inverter 7.69%

Table 2 Comparison

PARAMETERS	EXISTING SYSTEM	PROPOSED SYSTEM
Levels	5	15
No of source	2	3
No of switch	8	10
THD value	9.78	7.699

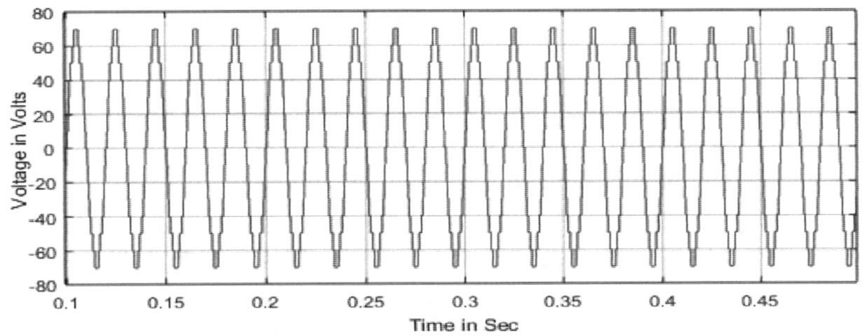

Figure 10. Level output voltage

Figure 11. Hardware

Figure 12. Hardware results

5. Conclusion

In this paper an asymmetric improved cascaded modified H-bridge asymmetrical multilevel inverter (CMHB-MLI) is proposed and generates a output voltage of 15 level, compared with the existing system with five level asymmetrical multilevel inverter the percentage of number of switches is reduced to 10, The analysis for total harmonic distortion (THD) in proposed CMHB-MLI by using Opposite phase disposition pulse width modulation (OPD-PWM) switching technique at lower frequency of (50 Hz) is

7.699% which is comparatively 40% less than existing system thereby reduces the switching stress and quality of the output improved without output filter, The experimental output is verified by using MATLAB/SIMULINK and hardware implementation.

References:

[1] C. H. Hsieh, T. J. Liang, S. M. Chen, and S. W. Tsai, "Design and Implementation of a Novel Multilevel DC-AC Inverter," *IEEE Trans. Ind. Appl.*, 2016, doi: 10.1109/TIA.2016.2527622.

[2] K.B. Bhaskar and T.S.Sivakumaran 'Implimentation of 11 level cascaded multilevel inverter using level shifting pulse with modulation technique with different loads', International journal of electrical engineering(IPASJ),volume 2,issue 10 Nov 2014

[3] L. Wang, Q. H. Wu, and W. Tang, "Novel Cascaded Switched-Diode Multilevel Inverter for Renewable Energy Integration," *IEEE Trans. Energy Convers.*, 2017, doi: 10.1109/TEC.2017.2710352.

[4] N. Prabaharan and K. Palanisamy, "A Single Phase Grid Connected Hybrid Multilevel Inverter for Interfacing Photo-voltaic System," 2016, doi: 10.1016/j.egypro.2016.11.281.

[5] Bhaskar, K. B., &Alagumariappan, P. (2020). Energy Channeling Led Driver Technology to achieve Flicker Free Operation with Zeta Converter for Power Quality Improvement. *International Journal of Automation and Smart Technology*, *10*(1), 401-407.

[6] S. Sudha Letha, T. Thakur, and J. Kumar, "Harmonic elimination of a photo-voltaic based cascaded H-bridge multilevel inverter using PSO (particle swarm optimization) for induction motor drive," *Energy*, 2016, doi: 10.1016/j.energy.2016.04.033.

[7] Paramasivam A., Bhaskar K.B., Madhanakkumar N., Vanchinathan C. (2020) Analysis of an Enhanced Positive Output Super-Lift Luo Converter for Renewable Energy Applications. In: Siano P., Jamuna K. (eds) Advances in Smart Grid Technology. Lecture Notes in Electrical Engineering, vol 687. Springer, Singapore. https://doi.org/10.1007/978-981-15-7245-6_11

[8] J. W. Dixon, "Multilevel converters," in *Power Electronic Converters and Systems: Frontiers and Applications*, 2016.

[9] L. K. Haw, M. S. A. Dahidah, and H. A. F. Almurib, "SHE-PWM cascaded multilevel inverter with adjustable DC voltage levels control for STATCOM applications," *IEEE Trans. Power Electron.*, 2014, doi: 10.1109/TPEL.2014.2306455.

[10] Alagumariappan, Paramasivam, Najumnissa Jamal Dewan, GughanNarasimhanMuthukrishnan, Bhaskar K. Bojji Raju, Ramzan Ali Arshad Bilal, and VijayalakshmiSankaran. 2020. "Intelligent Plant Disease Identification System Using Machine Learning" *Engineering Proceedings* 2, no. 1: 49. https://doi.org/10.3390/ecsa-7-08160

[11] S. T. Meraj, A. Ahmed, L. K. Haw, A. Arif, and A. Masaoud, "DSP Based Implementation of SHE-PWM for Cross-Switched Multilevel Inverter," 2019, doi: 10.1109/CSPA.2019.8695981.

[12] S. Bhadra and H. Patangia, "A comparison based harmonic elimination method with digital control of fundamental amplitude," 2013, doi: 10.1109/ICIInfS.2013.6731991.

[13] A. Prayag and S. Bodkhe, "A comparative analysis of classical three phase multilevel (five level) inverter topologies," 2017, doi: 10.1109/ICPEICES.2016.7853567.

[14] I. H. Shanono, N. R. H. Abdullah, and A. Muhammad, "A Survey of Multilevel Voltage Source Inverter Topologies, Controls, and Applications," *Int. J. Power Electron. Drive Syst.*, 2018, doi: 10.11591/ijpeds.v9.i3.pp1186-1201.

[15] X. Jiang and M. L. Doumbia, "Comparative Study of Grid-Connected Multilevel Inverters for

High Power Phtovoltaic Systems," 2019, doi: 10.1109/SEGE.2019.8859784.

[16] I. H. Shanono, N. R. H. Abdullah, and A. Muhammad, "A survey of multilevel voltage source inverter topologies, controls and applications," *International Journal of Power Electronics and Drive Systems*. 2018, doi: 10.11591/ijpeds.v9n3.pp1186-1201.

[17] M. S. Manoharan, A. Ahmed, H. W. Kim, and J. H. Park, "A single-source photovoltaic power conditioning system using asymmetric cascaded multilevel inverter," 2015, doi: 10.1109/ICPE.2015.7167761.

[18] A. Chen, D. Shao, C. Du, and C. Zhang, "High-frequency DC link flyback single phase inverter for grid-connected photovoltaic system," 2010, doi: 10.1109/PEDG.2010.5545796.

[19] K. H. Law, W. P. Q. Ng, and W. K. Wong, "Flyback cascaded multilevel inverter based SHE-PWM control for STATCOM applications," *Int. J. Power Electron. Drive Syst.*, 2017, doi: 10.11591/ijpeds.v8i1.pp100-108.

[20] A. Nami, F. Zare, A. Ghosh, and F. Blaabjerg, "A hybrid cascade converter topology with series-connected symmetrical and asymmetrical diode-clamped H-bridge cells," *IEEE Trans. Power Electron.*, 2011, doi: 10.1109/TPEL.2009.2031115.

[21] Y. He, P. Liu, J. Liu, and Z. Wang, "A design method of hybrid cascade multilevel structure for active power filter application in moderate-voltage grid," 2010, doi: 10.1109/IPEC.2010.5542339.

[22] S. Mariethoz and A. Rufer, "New configurations for the three-phase asymmetrical multilevel inverter," 2004, doi: 10.1109/ias.2004.1348509.

[23] Y. Suresh and A. K. Panda, "Investigation on hybrid cascaded multilevel inverter with reduced dc sources," *Renewable and Sustainable Energy Reviews*. 2013, doi: 10.1016/j.rser.2013.04.027.

[24] F. Khoucha, S. M. Lagoun, K. Marouani, A. Kheloui, and M. E. H. Benbouzid, "Hybrid cascaded H-bridge multilevel-inverter induction-motor-drive direct torque control for automotive applications," *IEEE Trans. Ind. Electron.*, 2010, doi: 10.1109/TIE.2009.2037105.

[25] Z. Du, L. M. Tolbert, and J. N. Chiasson, "Active harmonic elimination for multilevel converters," *IEEE Trans. Power Electron.*, 2006, doi: 10.1109/TPEL.2005.869757.

[26] J. Napoles *et al.*, "Selective harmonic mitigation technique for cascaded H-bridge converters with nonequal DC link voltages," *IEEE Trans. Ind. Electron.*, 2013, doi: 10.1109/TIE.2012.2192896.

[27] D. A. Ruiz-Caballero, R. M. Ramos-Astudillo, S. A. Mussa, and M. L. Heldwein, "Symmetrical hybrid multilevel DCAC converters with reduced number of insulated DC supplies," *IEEE Trans. Ind. Electron.*, 2010, doi: 10.1109/TIE.2009.2036636.

AUTHOR INDEX